RESEARCH REPORTS
OF THE LINK ENERGY FELLOWS

VOLUME 7

September 1991

RESEARCH REPORTS
OF THE
LINK ENERGY FELLOWS

VOLUME 7

September 1991

BRIAN J. THOMPSON, EDITOR

Published by
The University of Rochester Press
in association with
The Link Foundation

First published 1992

University of Rochester Press
200 Administration Building, University of Rochester
Rochester, New York 14627, USA
and at PO Box 9, Woodbridge, Suffolk IP12 3DF, UK

ISBN 1 878822 11 X

Library of Congress Cataloging-in-Publication Data applied for

British Library Cataloguing-in-Publication Data
A catalogue record for this book is available from the British Library

This publication is printed on acid-free paper

Printed in The United States of America

TABLE OF CONTENTS

Reports of the Link Energy Fellows, September 1991

THE LINK FOUNDATION
ENERGY FELLOWSHIP PROGRAM

The Link Foundation was established in 1953 by Mr and Mrs Edwin A. Link. The purpose of the Link Foundation is to promote the general welfare through the advancement of scientific, technological and general educational projects. The University of Rochester administers the Link Foundation Energy Fellowships. Objectives of this program are to foster energy research and to disseminate results of that research through lectures, seminars and publication. Since 1984 grants have been made annually to universities or other non-profit organizations which select Doctoral student fellows and supervising faculty or research directors to pursue an energy-related project. Ideas that can be implemented in the relatively near future are given priority by an independent selection committee.

THE LINK FOUNDATION

PREFACE

This volume contains a series of research papers written by the 1990–91 Link Foundation Energy Fellows. Each of the fellows is working towards a Ph.D. degree and the research reported is part of the thesis work conducted during their tenure as fellows. The Link Foundation is pleased to make these research papers available to other workers in the field of energy studies and also to provide a record of the the accomplishments of these outstanding students. This is the seventh volume in a series of publication and again represents a diverse set of important contributions to the literature.

The Link Fellowship Program is funded by the Link Foundation to both honour and memorialize Edwin A. Link. The Link Foundation was established in 1953 by Mr and Mrs Edwin A. Link. It is the policy of the Foundation to make grants to qualified non-profit organizations interested in the mastery of the air and sea, and the development of energy resources and their conservation. Each year fellowships have been awarded in these categories as a means of implementing this policy.

The University of Rochester has the responsibility to administer, on behalf of the Link Foundation, the Link Energy Fellows Program.

The fellows, whose work is reported here, were selected from a large number of applications in a competitive process by an independent review committee. The committee members and the Foundation are convinced that the fellows represent some of the best young scholars in the energy and energy related fields. We are proud of the success of these students.

Brian J. Thompson

SELECTION COMMITTEE FOR 1991 FELLOWS

Provost **Brian J. Thompson** *Chair*
University of Rochester
Rochester, New York

Dean **Bruce W. Arden**
University of Rochester
Rochester, New York

Professor **William N. Gill**
Rensselaer Polytechnic Institute
Troy, New York

Provost **Angel G. Jordan**
Carnegie Mellon University
Pittsburgh, Pennsylvania

Dr **Irvin L. White**
New York State Energy Research and
Development Authority
Albany, New York

Professor **Lee Lynd**
Dartmouth College
Hanover, New Hampshire

An Engineering-Economic Evaluation of Offshore Oil and Gas Development in the Ross Sea, Antarctica

DIETER BEIKE

Petroleum Engineering Department
The University of Texas at Austin
Research Supervisor: Dr W.C.J. van Rensburg

ABSTRACT

This paper speculates about the viability of Antarctic petroleum development and estimates when Antarctic oil production will become economically feasible. This analysis is based on a hypothetical exploration and development model which could require a US $7.6 billion investment in 1991 constant dollars. Following a 20 year development period, a 30 year project life, producing an initial 144,000,000 bbl per year, with a 5% decline rate is assumed. Exploration is based on a floating conical drilling unit. Production is based on a concrete Tension Leg Platform concept, with subsea completion. Storage will be in a seafloor facility and transportation will be by icebreaking tankers.

The Ross Sea was chosen as the most promising petroleum-producing area in the Antarctic, because it has the most promising sedimentary basins, has had the most scientific exposure, and is environmentally the most favorable area.

A NPV of US $844,762 at a real discount rate of 10% and an IRR of less than 10.01%, was calculated by using an oil price of $50/bbl (1991 $US). However, in the event of an oil spill costing US $2.5 billion (constant $) in year 9, an oil price of US $76 would be required to make this a viable operation. At a current (1991) oil price of $20/bbl, a production of 360,000,000 bbls per year, declining at 5% per year would be required, or capital costs would have to decline by 70%, or operating costs would have to be negative in order for this project to survive.

The risk of this venture, with its large upfront investment, spread over 20 years, makes it unlikely that such an operation will be economically feasible.

INTRODUCTION

Over the past 15 years increased speculation on hypothetical Antarctic oil and gas resources has led to an increasing number of publications dealing with the potential mineral resource of this continent. The emerging importance of an Antarctic mineral regime within the Antarctic Treaty has been recognized by the international community. The hypothetical wealth of this resource base interested the administrative-legal and bureaucratic levels of governments.

An assessment of the existing knowledge of possible oil and gas reservoirs is the basis of a hypothetical exploration and production scheme. This is followed by an assessment of the environmental impact of oil pollution in this fragile environment. Potential cost figures are derived from proposed operations in the Arctic and are used as a basis for an economic evaluation of such a venture. One earlier study by the Office of Technology Assessment[1] outlined the possible development of such an operation, assuming that $28.3 billion in capital costs and $2.3 billion annually in operating costs would have to be spent to produce oil at a price of $40/bbl in a 4 billion barrel oil field.

Since Antarctica is a truly international area, its resource development will have a substantial geopolitical impact on governmental and non-governmental activities. The Antarctic Treaty, which was ratified in 1961, regulates all activities on this continent. However, mineral resources are not covered by this treaty. On July 4, 1991 President Bush announced that the United States would join the other Antarctic Treaty partners in a provision that specifies that all mineral development activities concerning Antarctica will be subject to a 50 year moratorium. Thus, the ratification of a mineral regime, which was negotiated over the past 10 years, failed. Such a regime is debated. Proponents of this regime believe that it will enhance the effectiveness of the Antarctic Treaty and protect the Antarctic environment from unregulated development which might lead to severe disturbances of the ecological system. Opponents believe, that such a regulation would open Antarctica for mineral development and then naturally damage the pristine environment. Therefore, the moratorium agreement prevents a mineral regime.

GEOLOGY AND MINERAL RESOURCES

Regional Geology

Antarctica is part of the Antarctic plate that extends to the mid-oceanic ridges in the Pacific, Indian and South Atlantic Oceans. Two geologically contrasting provinces form the Antarctic continent. A Precambrian Craton flanked by a Lower Paleozoic orogenic belt (Transantarctic Mountains), which is overlain by flatlying Upper Paleozoic to Lower Mesozoic sedimentary rocks, called the Beacon Supergroup, and intrusive as well as extrusive rocks of the

Ferrar Group.[2] West Antarctica, on the contrary, consists of a series of orogenic belts of Early Paleozoic to Late Mesozoic–Early Cenozoic age and probably consists of several microplates[3] where exposed rocks are mainly of Mesozoic and Cenozoic age.

Antarctica was part of the prehistoric supercontinent Gondwana, which existed between Late Precambrian through Early Jurassic time. The relationship between South America and Africa, India and East Antarctica, as well as East Antarctica and Australia are understood to a satisfying degree. A major disagreement remains concerning the eastern and western parts of Gondwanaland and the Paleo-Pacific margin which centers mainly on West Antarctica and the Antarctic Peninsula-Falkland Plateau overlap.[3]

Reconstructions and correlations of Antarctica in reference to the other Gondwana continents were largely based on morphology, paleomagnetism, matching of magnetic anomalies in ocean basins, and far less on matching of rocks or structures between continents.[4] Craddock[5] cites as evidence of Antarctica's role in Gondwana the matching of basement rocks of Precambrian-Mesozoic age, distribution of Upper Paleozoic glacial deposits, Paleozoic-Mesozoic fossil records, Antarctic orogenies with continuations in other continents, widespread Jurassic basaltic volcanic rocks, magnetic anomaly belts in the oceanic regions, and the morphology of continental margins.

The East Antarctic shield is a Precambrian craton consisting of high-grade metamorphic rocks and silicic to mafic plutonic rocks.[4] The stratigraphy of the shield is not very well known; the oldest known rocks are determined to be 3.8 billion years old. According to James and Tingey,[6] the shield is composed of Proterozoic rocks, including a substantial amount of Archean rocks. Crystalline rocks of Precambrian age, forming the igneous and metamorphic shield, are the major rock type between 2° West and 150° East.[7]

West Antarctica's continental crust is much thinner than East Antarctica's and thus it is topographically lower. Rocks of West Antarctica are usually of younger age and are mostly related to an active convergent Pacific margin belt and subduction of the oceanic crust during and after the Gondwana breakup.

Late Paleozoic and Early Mesozoic rocks are very common in West Antarctica, which occur extensively in the form of outcrops of thick greywacke-shale sequences, including locally shallow water marine sediments with Triassic sediments.[2] An active plate margin is demonstrated by the geologic record of that time. Cenozoic time records show that oceanic crust was subducted beneath the Pacific rim of Antarctica and thus is related to the seafloor anomaly pattern of the southeastern Pacific.[2] According to Elliot,[2] the Late Mesozoic to Early Cenozoic was characterized by diminishing land connections in the southern hemisphere, leaving behind, after the initial separation of Antarctica and South America, microplates of West Antarctica as land connections. Rowley[4] identified four large continental plates that contain exposed rock – Ellsworth Mountains, Antarctic Peninsula–eastern Ellsworth Land area, western Ellsworth Land, and Marie Byrd Land. However, rocks and structural trends seem to differ. Most of the larger and smaller plates resulted from fragmentation of the continental lithosphere on the Pacific side of present East Antarctica. It appears, that no long migration path of these plates can be recorded.

Petroleum Potential of Antarctica

The tectonic history of Antarctica subsequent to the Gondwana breakup suggests that West Antarctica is the most likely area to contain petroleum resources, although the potential for petroleum accumulation also exists in East Antarctica.[8]

St John[9] estimated a potential hydrocarbon yield of 203 billion barrels (about 30 billion tons). This was based on a factor of about 1700 tons of hydrocarbon per cubic kilometer of sedimentary rock. Behrendt[10] dismissed this value as unrealistically high, since St. John did not have available published interpretations of the seismic reflection data to calculate sedimentary rock thickness, due to compilation of his data in 1980. The USGS[11] estimated 19 billion barrels (about 3 billion tons) at the 5% confidence level, which is considered unrealistically low by Behrendt.[10]

According to the USGS,[11] Antarctica must possess the same limited petroleum geology as its pre-breakup neighbors, southern South America, South Africa, India, and Australia. Moderately large petroleum occurrences would require a craton-margin rift development or a delta. The former seems to be a distinct possibility (as an analogue to the Gippsland Basin in southeastern Australia), while the latter seems less likely due to the Late Neogene history in Antarctica of glacial cover. However, the time between breakup and glaciation could have led to delta occurrence[11] Possible areas for petroleum potential in the Antarctic include the Wilkes Land Sector, Queen Maude Land/Amery Sector, the Weddell Sea Sector, the Antarctic Peninsula region, and the Bransfield Basin. This report focuses on the Ross Sea, in particular the Victoria Land Basin.

The Ross Sea

Three sedimentary basins in the Ross Sea were identified by Davey, Bennett and Houtz.[12] These are the Eastern Basin, the Central Trough, and the Victoria Land Basin (VLB). These basins received sediments since at least the Oligocene.[13] The Ross Basins cover 1,627 million km^2 and are estimated to contain 3.82 million km^3 of sediment. The potential hydrocarbon yield is calculated to be 45.8 billion BOE.[9] Houtz and Davey[14] described sediments of up to 4 km in thickness beneath the eastern shelf and gentle synclines and anticlines under the western part of the shelf. Behrendt[15] interpreted data from beneath the Ross Ice Shelf to show sedimentary sequences up to 8 km thick. All three depocenters are underlain by large basement grabens that are filled with rift-related sediment of 3–8 km thickness. These early-rift grabens are buried beneath a widespread glacial sedimentary section 2–6 km thick.[16] The widths of these large early-rift grabens define the widths of the VLB and the Central Trough, whereas the early-rift graben of the Eastern basin covers only a small part of the basin.[16] These early-rift grabens are part of a complex rift-zone that formed during the Mesozoic breakup of Gondwana and once stretched several thousand kilometers from Australia, New Zealand and across Antarctica through the Ross Embayment along the Transantarctic Mountains to the Weddell Sea.[16] Active rifting in the Ross Embayment is presently only identified in the VLB, Marie Byrd Land and possibly along the Central High.[16]

Two phases of rifting have been proposed for the Ross embayment region,[17] early rifting from late Jurassic/Early Cretaceous to Late Creataceous, and a late rift phase from the Eocene to the present. Initial uplift, erosion and deformation of Beacon Group Strata and intrusion and extrusion of Ferrar dolerite sills and Kirkpatrick basalt is thought to be caused by crustal extension during the early rift period. This was then followed by rapid downfaulting of the major early-rift grabens which were then infilled with marine and non-marine sediments during late Cretaceous and Paleogene time. The late rift period is characterised by uplift of the Transantarctic Mountains, development of the Terror Rift and volcanism in Marie Byrd Land.

Fritsch[18] reported two discontinuities in the eastern part of the Ross Sea continental shelf which he correlated with upper Miocene-lower Pliocene and Middle-Miocene. Upper Miocene contacts were detected in DSDP coreholes. These DSDP drillholes (leg 28 holes 270, 271, 272, and 273) indicated a Paleozoic continental basement overlain by Oligocene, Miocene, Pliocene and Pleistocene rocks. Presence of littoral glauconitic and carbonaceous sands indicate a relatively temperate environment before the Late Oligocene.[19] Amounts of methane and ethane were detected[20] and interpreted by McIver[21] as originating from local organic diagenesis. Ethane could have accumulate due to its migration from deeply buried rocks undergoing early stages of thermal cracking, leading to the possible generation of light hydrocarbons. Claypool and Kvenvolden,[22] however, suggest, that the ethane could be generated also by microbiological processes. They reported high isotopic delta-13 carbon values of −78.9 to −67.5 for the methane gas, indicating possible biologic origin. Alternatively, it could be assumed, that the gases have been concentrated at shallow depths as gas hydrates,[21] however, a bottom simulating reflector (BSR), that commonly indicates gas hydrates, has not been reported from the Ross Sea.[23] First direct evidence of liquid hydrocarbon generation in the Ross Sea was found by the CIROS–1 drillhole where asphaltic residue was discovered in a 2m section 632–634 m below seafloor (mbsf) 28 (see below VLB).

The Ross West Basin is possibly similar in structure and stratigraphy to the productive Gippsland Basin of southern Australia, whereas the Ross East Basin may prove comparable to the basins of the Campbell Plateau offshore New Zealand,[9] which had proved reserves of 345 million tons (2.5 billion bbl) of oil and 220 billion m^3 of gas.[24]

According to Lock,[13] the Ross Sea Region sedimentary sequences deposited during continental breakup show favorable structural and facies relationships with depositional environments, including coastal plains, lagoons, swamps, deltas, and continental slope fans.

In general, petroleum producing provinces of southern Gondwana are related to passive margins and usually produce paraffin-based crude oils and some gas. The Gippsland Basin is an example of these waxy crudes, that commonly are associated with humic type kerogen derived from terrestrial flora. Conifer species existed during the Jurassic and it can be expected that they were well distributed throughout Gondwana and its final breakup. Auracian conifers played a major role in the organic development of the Gippsland Basin. These conifers are the producers of thick waxy matter and hydrogen rich resins that are resistant to bacterial degradation.[25] McIntyre and Wilson[26] recorded fossil pollens from these plants in Antarctic lower Tertiary erratics and samples from DSDP site 270 show recycled pollens. Deposition in cold acid peat swamps would have preserved much of this oil generative humic matter.[13] However, according to Cameron,[27] comparisons have to be looked at very carefully, as the Gippsland Basin is not part of the southern Australian marginal rift and therefore has no direct analog on the Antarctic margin.

Petroleum Geology of the Victoria Land Basin

The area covered by the Victoria Land Basin (VLB) is about 45,000 km^2. It is the westernmost of the three Ross Sea basins. The basin stretches in a north-south direction along the Transantarctic Mountains (in the west) and the submarine Coulman High (in the east) from Cape Washington (northern Victoria Land) to the Ross Ice Shelf, with its southern margin being unknown.[28]

Glacial sedimentary rocks comprise a significant percentage of the strata that cover the Ross Sea.[19] Such glacial strata has no source rock potential. Hydrocarbon potential will be poor, if they persist to great depth.[16] Hydrocarbon potential, therefore, is heavily tied to preglacial deposition of sediments that are deeply buried within the rift grabens. No such preglacial sections have been sampled in situ in the Ross Sea. Therefore, speculation of possible hydrocarbon accumulations in the Ross Sea are speculative, although conditions favorable for hydrocarbon generation and entrapment are likely in the Ross Sea – particularly in the VLB, provided that adequate source beds exist.[16] Table 1 summarizes current knowledge of the petroleum geology.

Hydrocarbon Generating Windows and Petroleum Potential

Three models for hydrocarbon generating windows have been developed.[29,16,30] The hydrocarbon generation zone was assumed to be at 2.5 to 4.0 km depth, varying by less than 600 m (vertically) between depocenters.[16] Temperature changes affect oil window depth more than shifts in relative time of sedimentation and subsidence. Present oil window locations near the edges of depocenters is uncertain. In case unconformity U6 serves as the base of Cenozoic glacial deposits, then the position of the oil window above or below U6 is important to possible present day hydrocarbon generation because source beds are not likely in glacial deposits.[16] Hydrocarbons probably were generated in Cenozoic and Mesozoic times within the thickness of rift-related rocks that

Table 1. Petroleum Geology of the Victoria Land Basin

Structure
- normal faulting with large offset basement faults, horst and grabens with probable intrusives exists close to the coast (western margin of basin);[104,105]
- late Precambrian to early Paleozoic meta-sedimentary and granitic rocks overlain by flatlying non-marine sedimentary strata of Devonian to Early Jurassic age are intruded by Middle Jurassic sills of dolerite underlie the VLB;[28,106]
- an active rift zone (Terror Rift) comprising the sediment filled Discovery Graben and the adjacent magmatically intruded Lee Arch run in a North-South direction, superimposed upon older basin structures and deeply buried strata;[28]
- these early rift grabens originated possibly during Middle Jurassic to early Cretaceous as part of the Gondwana breakup; a younger phase of rifting and sedimentation during Cretaceous to early Tertiary with subsequent deformation and uplift of the Transantarctic Mountains is also possible[104,105]

Stratigraphy
- the upper part of the sequence (26–366m) was dominated by diamictite (40% – formed by mud, sand and stones) and the lower part (366m–702m) was dominated by largely deep water mudstone with sandstone beds and occasional conglomerate deposited from gravity flows[28] (CIROS–1 drillhole)

Source rocks
- total organic carbon (TOC) values are low and average 0.34%,[107] however average for all Ross Sea drillholes is 0.37%;[16]
- high oxygen indices (45–1942 mg/g) TOC and very low hydrogen indices (3–24 mg/g TOC) are indicated by Pyrolysis;[107]
- this shows similarity to type III kerogen classified by Tissot et al.;[108] much of the particulate matter is present as reworked coal;[32]
- the hydrocarbon generating potential of the samples is low (0.06–0.34 mg hydrocarbon/g of rock;[28]
- abundance of marine microorganisms was reported from CIROS–1;[28]
- presence and impact of terrestial organic matter is possible due to probable existence of coastal forests near the drillsite in the Oligocene[109] and supported by a beech leaf remnant in the core of the drillsite[110]

Organic maturation
- Tmax ranges from 393 – 422°C;[107]
- a mean max. vitrinite reflectance of .36 (0.03%)[32] was measured indicating that the remaining sediments are not thermally mature for hydrocarbon generation;[28]
- maturity decreased with depth and the production ratio S1/(S1+S2) decreased with depth;[107]
- the vitrinite reflectance for MSSTS–1 was measured as less or equal to 0.54%;[111]

Temperature	– temperature relations are poorly understood; Ross Sea subsurface temperatures are presently and have been above normal from the Mesozoic to the present due to extensive crustal rifting; heatflow values are generally high, indicating that the entire Ross Sea region is tectonically active; a temperature gradient of 40°C/km was reported for CIROS–1;[112]
	– a shallow crustal deformation is currently concentrated along the VLB and Marie Byrd Land;[16]
	– a three stage thermal subsidence history was proposed[16]
	   Stage 1: regional crustal stretching and downfaulting of early rift grabens,
	   Stage 2: crustal subsidence by cooling,
	   Stage 3: renewed crustal stretching and downfaulting (late rift stage, which is confined to the eastern and western edges of the Ross Sea);
	– the temperature depth histories are suitable for generation of hydrocarbons at subseafloor depths of 2.5–4 km (TTI = 15 – 160)[16]

Reservoirs	– permeability measurements do not exist; – visual porosity ranges from 0–15%[113] and corrected log porosities of up to 21.6%[112] were reported – good secondary porosity is present in a number of sandstones resulting from dissolution of carbonate cement, bioclastic debris and feldspar;[28] – based on analogy with rocks drilled from coeval ains on the Campbell plateau and western Tasmania[114] likely Mesozoic to Oligocene preglacial rocks may contain reservoir rocks in deeply buried sandstones, which might lead to poor quality reservoirs due to deep burial, cementation and thermally controlled diagenesis[16]
Traps	– most likely at great depth near basement faults and at all depths within the Terror Rift and along the westside of the VLB;[16] – stratigraphic traps are likely and could occur at great depth along the edges of the early rift grabens and at shallower depth within the divergent sections cut by unconformities on the edges of a depocenter[16]
Migration	– hydrocarbons migrated probably laterally along dipping strata that are common at the edges of depocenters;[30] – basinward dipping strata with numerous unconformities are found along the entire western edge of the VLB;[28] – vertical migration below U6 is possible;[16] – direct pathways due to late rift fault zones or intrusive structures could exist in the Terror Rift of the VLB, possibly providing for upward migration from deeply buried sites to shallower sites[28]

lie below the present oil window and have previously reached peak hydrocarbon generation.[16]

Petroleum potential of the VLB: Sediments in the VLB drilled to date have low organic content. Kerogens are mainly derived from reworked, oxidized organic material. Generating potential for hydrocarbons is very low to negligible with low values for organic maturity.[28] Residual oil saturation in the CIROS–1 drillsite is very low. A dominantly terrestrial source is suggested[31] and Lowery[32] estimated that reflectances of 0.5 to 0.6% could only be attained by sediments now at depths of at least 1500 to 2000 m. Source of the residual oil is therefore most likely to be sediments in more distal parts of the basin that contain richer, less oxidized and possibly marine organic matter and have been within the oil window.[28]

Rapp et al.[33] inferred from near-surface sediments that important inputs from terrestrial plants can be assumed and Sackett et al.[34] suggested that up to 90% of the organic matter in Ross Sea sediments is reworked. Thus organic sedimentation reported for the CIROS–1 drillhole might be representative for the VLB since early Oligocene and hence oil-generating capacity in source rocks of this age might be limited over the whole basin.[28] However, Waples-Lopatin models for the VLB[29,16] indicate that the lowermost sediments in the basin are within the oil generation zone and that the basin has good modeled potential. The probable organic contents inferred by Cook and Davey,[29] however, were substantially higher than the actually measured values from drill holes and a time-temperature index of 1529,[16] was used to predict the beginning of hydrocarbon generation and the top of the oil window. Terrestrially derived kerogen, however, might require higher temperature values, as shown by Cook[35] and Johnston et al.[36] These studies showed for western basins of New Zealand necessary vitrinite reflectance values of at least 1.0% for terrestrial organic matter to be released as oils from the source rock.

Collen and Barrett[28] summarize CIROS–1 data in proclaiming that unfavorable source characteristics exist in the Miocene to early Oligocene sediments. Only minute amounts of residual oil was found in sandstones with good porosity and seals. They assess the possible hydrocarbon prospect of the glacial sequence of the basin in the McMurdo Sound area as low.

THE MARINE ENVIRONMENT

The Southern Ocean surrounding Antarctica represent a unique regional marine environment. These waters cover about 36 million km^2 (10% of the world's oceans – with the Ross Sea covering 200,000 km^2). These waters are characterised by vigorous zonal, meridional and vertical circulation. The most important hydrographic boundaries include the Polar Front Zone (Antarctic Convergence) between 47°S and 63°S, where cold Antarctic surface water sinks below warmer sub-Antarctic water and the Antarctic Divergence at 60°S to 63°S, which is induced by east and west winds, causing surface water to be transported horizontally, which then in turn is replaced by colder Circumpolar Deep Water due to an upwelling mechanism.

Sea ice conditions are characterized by a large seasonal change, with a minimum extent of 3 million km^2 in February to a maximum of 21 million km^2 in September. The duration of the open water season is highly variable. Break-up of the ice usually starts in November and freeze-up starts in late February-early March, depending on the development of reoccurring polynyas.

Icebergs are produced in greatest volume from ice shelves, followed by ice streams and active outlet glaciers. These bergs drift in erratic patterns and reach their northern limit usually at the Polar Front Zone.

Meteorologically the area is characterized by a continous darkness (light) for 2.5 to 4 months in the winter (summer). Frequent storms are present between 60°S and 70°S usually travelling in eastward direction. Strong and persistent katabatic winds are predominant on the coast. Between February and May is the windiest period of the year. The area experiences one of the cloudiest zones on earth and low visibility occurs about 10–20 % of the time in Ross Sea. Temperatures are very low and can reach winter extreme minima in the minus sixties.

Chemically, the southern ocean consists of very dense waters, indicated by high salinities.

Biologically the Southern Ocean is a highly productive area mainly consisting of phytoplankton, krill, squid, fish, penguins, sea birds, seals and whales. The foodweb is short, with few species but large numbers of individuals, with krill being the primary link between primary consumers and higher trophic levels. Primary productivity and biomass is seasonally highly concentrated between early summer and late fall.

Seasonality of light conditions, therefore, are the driving force behind the Southern Ocean ecosystem. Limitations in the visible part of the electromagnetic spectrum reduces the energy flow through the ecological system to a short intense season. Biogeochemical cycles are not limited as indicated by high nutrient levels in the Ross Sea, and do not limit biological productivity. This seasonality leads to a less stratified thermocline, which is evidenced by higher salinities and a reduced halocline.

ENGINEERING CONSIDERATIONS

Environmental Design Criteria

The Ross Sea environment is extremely severe in comparison to other areas for offshore development. Capital and operating costs for offshore petroleum development are directly related to the environmental constraints placed upon these systems. The current limited knowledge of Antarctic environmental conditions has been compiled by Keys.[39] Table 2 on environmental design criteria is compiled from that publication, in particular Appendix 1, 2, and 3.

Table 2. Oceanographic and Meteorological Characteristics
(compiled after source 39 – in particular Appendix 1, 2 and 3)

Phenomenon	Description	Comments
BATHYMETRY		
water depth:	– 200–900m	
TOPOGRAPHY		
seafloor topography	broad ridges – north-south trending and seperated by troughs	
land topography	– Transantarctic Mountains (steep and high relief)	– area of katabatic wind phenomena
PHYSICAL OCEANOGRAPHY		
greatest medium wave height	– ca. 2m	
max. significant wave height	– at least 5–10m	
wave height (Jan.– March)	– < 1 m (western Ross Sea) 2–4 m (avg. annual at 60°S)	
max. wave period	– 16 seconds	
tides	– diurnal	
tidal flow	– reaches 5 knots at Victoria Land coast	
max. spring tidal range	– ca. 2m	
west-setting currents	– 1–3 knots (along Ross Ice shelf – southern limb of gyre)	– 1. water masses drift towards Victoria Land 2. major exit for water masses is north-west Ross Sea
tsunamis	– low risks	
bottom currents	– 0.2 knots	
max. seasurface temperature	– 0°C	

Phenomenon	Description	Comments
CHEMICAL OCEANOGRAPHY		
Antarctic Surface Water	– 0 to minus 1.9°C 34 to 35 ppt salinity	
warm Circumpolar Deep Water	– modified in Ross Sea to warmer than –1°C	
cold saline Ross Sea Shelf Water	– minus 1.8 to – 1.95°C > 34.75 ppt salinity	
very cold ice water	– less than –2°C	
surface seawater densities	– 1027.4 to 1028 kg/m^3	
METEOROLOGY		
mean annual air temperature at adjacent land stations	– minus 21 to –14°C	
mean monthly air temperature at adjacent land stations	– minus 31 to 0°C	
warmest mean annual temperature over Ross Sea	– minus 10°C	
approx. mean annual wind speed	– minus 3–10 m/sec	
mean mid-winter wind chill values at adjacent land stations (Jan – July)	– 1400–2200 watts/m^2	
approx. max. gust	– 51 m/sec	
mean annual wind speed	– 3–10 m/sec	
storms with consistent wind speed of approx. 30 knots	– can last up to 5 days	
mean monthly cloud cover	– 4/10 (winter) to 9/10 (summer)	– aircraft icing common
cloud ceiling	– often below 600 m (western Ross Sea)	
low visibility	– 10–20 % of total time	

Phenomenon	Description	Comments
ICE CONDITIONS		
sea ice type	– mainly dynamic first-year pack ice	
sea ice thickness	– max. undeformed thickness > 2m	
max. ice concentration	– 10 tenths	
average proportion of multiyear ice at midwinter	– 0 percent	
main type of ice movement	– divergent	
max. rate of sea ice movement–	– > 5 km/hr	
duration of open water season (months with < 1/10 ice)	– 0–5	– begins in November – ends late February
first year pressure ridge heights	– 4–5 m (western McMurdo Sound)	
ridge frequency	– < 8/ km (average)	
persistent pack ice	– north-eastern Ross Sea, western coast of Victoria Land	– due to East Wind Drift and Ross Sea gyre
shore-fast sea ice	–20–50 km wide off Victoria Land	
ICEBERG CONDITIONS		
mean annual iceberg production		
total mass (tonnes)	– $10\text{–}16 \times 10^{11}$	
total volume	– $1200\text{–}2000 \ km^3$	
total numbers (rough estimate)	– 6000–70 000	
mean density	– $840 \ kg \ m^3$	
estimated median mass	– $10^5\text{–}10^7$ tonnes	
max. recorded thickness	– 390 m	
max. recorded height	– 140 m	

Phenomenon	Description	Comments
most common berg width	– < 200 m	
median width of all icebergs (at waterline)	– < 100 m	
max. recorded draft	– 330 m	
avg. draft of tabular icebergs	– 100–200 m	
estimated max. mass	– 10^{12} tonnes	
max. concentration (bergs per 1000 km^2, i.e. within 17.8 km of observation point)	– 90	
in Ross Sea prospects at least 20 km offshore	– > 20	
max. recorded daily drift	– 46 km	
avg. measured speed	– 0.05 to 0.15 m/sec.	
max. ice draft (depth of zone subject to iceberg scour	– > 300 m	
GEOTECHNICAL CONDITIONS		
onshore		
coastline	– only 5% consists of rock and much of this is very steep	
beaches	– 3% of the total coast of western Ross Sea	
coasts with exposed rock	– 28% of the Ross Sea coastline	
low rocky shores or boulder/ gravel/sand beaches	– 7%	– suitable for shore-based facilities; penguin rookeries cover 50% of this area
risk of volcanic disturbances	– low	

Phenomenon	Description	Comments
permafrost	– up to 500 m thick on land around McMurdo Sound	
seasonally thawing active layer	– 0.5 to 1 m thick	
soil moisture in upper 0.3 m	– varies from 1 to 22%	
glacial meltwater (Dec.–Feb.)	– stream flow rates up to 1 m^3/sec	
offshore		
sediment	– poorly sorted glacial-marine; coarse sediments (sands, gravel) close to shore; fine-grained sediments common in offshore deeper waters	
surface deposits	– water saturated glacial marine	
substrate conditions	– cohesive diamictons	
differential compaction	– overcompacted orthotills	– possibly limited load bearing capacities; vibrational loads and added load of a platform might destabilize and mobilize glacial deposits[115]
interbedded deposits	– upper few hundred meters	
sedimentation rates	– very low in Ross Sea	
ice-cemented permafrost	– no firm evidence, water in upper 50 m of unfrozen sediments, sediments mostly soft and not indurated	
sub-seafloor cemented permafrost	– possibly as a relic from last ice age	
shallow gas	– encountered in four drillholes; pressures were not measured	
seabed slope	– 2°	

Phenomenon	Description	Comments
seafloor instability	– sediment gravity flow deposits very common	
clathrates	– can be expected just below seafloor at depth > 250 m	
faulting		
BIOLOGICAL OCEANOGRAPHY		
food web	– short (few species but large numbers of individuals)	
primary production	– strong seasonal fluctuation with peak between early summer and late autumn	
main primary producers	– drifting phytoplankton (diatoms, dinoflagellates, silico-flagellates)	
major link in foodweb	– krill	
photosynthetic activity	– highest in western and southern Ross Sea	
zooplankton	– mainly in upper 500 m of water column and drift with currents	
benthic fauna	– abundant and great variety of species; concentrated (between 30 and 500 m where hard substrates are exposed)	
benthic fauna	– marked horizontal and vertical gradients in McMurdo Sound slow growing benthos is believed to be valuable food source year round	
seabird breeding rookeries	– concentrated in western Ross Sea, mainly along the Victoria Land coastline north of Drygalski Ice Tongue and on Ross Island	
penguin breeding population	– 1/3 of total Antarctic population are in Ross Sea for two species: 700,000 pairs of Adelies, 40,000 pairs of Emperors	

Phenomenon	Description	Comments
number of individual rookeries	– 32–35	
concentration of seabirds	– highest in summer in rookeries	
activity radius of seabirds	– Adelie penguins active at sea within 300 km radius of rookeries	
stability of breeding population	– for Adelie population stable within 10%	
petrel population	– in the millions	
skua population	– in the thousands	
ecologically sensitive areas	– south-west and north-west Ross Sea, due to concentration of sea bird breeding rookeries; four small coastal areas in western Ross Sea with sea bird colonies or vegetated areas have been given protected status by the Antarctic Treaty	

LOGISTICS

Phenomenon	Description	Comments
max. distance of drillsites offshore	– 400 km	
duration of winter sunless period	– 2.5 (72°S) – 4 months (78°S)	
duration of summer continous daylight	– 2.5 (72°S) – 4 months (78°S)	
distance to next supply base (New Zealand)	– 2500 km	

The impact of these physical parameters on Antarctic offshore operation is shown in table 3.

Characteristic wind features: Cyclonic storms, temperature inversions and the katabatic wind phenomena are typical for Antarctic weather. A low pressure belt is formed between 60°S and 70°C latitude due to the mixing of cold Antarctic air with warmer air from lower latitudes. This results in various low pressure centers: one is located at 65°S to 75°S, 155° to 180°W north-east of Ross Sea, resulting in storms that can travel up to 2,000 km/day.[39,40]

Low-level temperature inversions are common in coastal areas and in the interior of the continent. It is caused by intense radiational cooling at the surface. Calm, clear conditions enhance inversions, resulting in average strengths exceeding 10°C in south-east Ross Sea during June – August.[41]

Katabatic winds are related to inversion strengths and steepness of the terrain.[42] Gravitational sinking of cooled dense surface air, flowing outwards from the interior of the continent towards the coast and then turning west – in the case of Victoria Land north – due to the Earth's rotation.[41] These winds are channelled into narrower, steeper regions where strong turbulent winds are persistent.[42] These winds diminish 10–40 km offshore but this range might be extended by sea ice.[39,41] The major Ross Sea surface wind features were summarized by Keys[39] Strong clockwise circulation occurs in the south and west, with strongest winds from the southerly quarter but prevailing winds from the south-easterly quarter. The surface wind field is influenced by topographic features. Highest mean monthly wind speeds tend to occur in the fall, but stronger winds with increased frequency of calms tend to occur in winter. Storms can occur any time of the year with their greatest frequency and duration in winter and spring.

Water Circulation: Large scale circulation of the Southern Ocean can be divided into zonal components – Antarctic Circumpolar Current (West Wind Drift) and Antarctic Coastal Current (East Wind Drift). This is superimposed by the meridional (north-south) components. Surface speeds of the Antarctic Circumpolar Current (located between 50° and 60°S) average 0.1–0.5 m/s with speeds at the seafloor of 10–40% of the average surface speed.[39] It is the largest current on earth, which is weakly stratified and has a water mass transport of 140 million m³/s eastward.[43] The East Wind Drift moves south of 65°S in a westward direction, averaging 0.05 to 0.5 m/s but can attain 1.5 m/s in the Ross Sea Gyre. The greatest meridional flow is northward and sinks at the Polar Front beneath Sub-Antarctic Surface water.

Surface water circulation is controlled by the Ross Sea Gyre. Westward current flow of 0.5 m/s – 1.5 m/s along the Ross Sea Ice Shelf and then northwards along Victoria Land of 0.3 m/s[44,45] is probably followed by an eastward flow just south of the continental shelf break to complete the gyre.[46] Keys[39] concluded from the presence of the gyre, that much surface water in the

Table 3. Environmental Forces and Impact on Offshore Operations

Phenomenon	Description	Impact on operational phase or component
wind	– act over wide areas (several hundred of miles) and persist for several days at hurricane force	– superstructure of a platform
waves	– breaking waves in deep water exert a very high local force of short duration and high intensity – presence of pack ice reduces the wave height, so that the resulting impact is reduced	– superstructure of a platform
currents	– create horizontal pressures against structural surfaces	– create vortex shedding on risers
ice	– global loads consist of the total force that an ice feature exerts on a structure, and are determined by limit stress load, limit momentum load, and limit force load – local loads are due to the fact that ice does not fail in a uniform pressure across the complete contact area between the structure and the ice feature and higher than average pressures over smaller areas can be exerted locally on the structure[5]	– towing a 3 mile long streamer for multichannel seismic data acquisition – increase of downtime during drilling operations – superstructure of a platform – storage and transportation layout
dynamic ice forces	– the force of impact is of an uneven, dynamic nature containing random and continously repetitive fluctuations[8] depending on environmental driving forces – single ice feature travelling at a high velocity will cause short term transient loading	– same as impact of ice – possible scouring of seabottom, therefore affecting seafloor installations
waterdepth	– forces on mooring lines increases	– stationkeeping of structure becomes very difficult
soil conditions	– unevenly distributed strength characteristics of the soil	– foundation design, trenching of pipelines, and burial of wellheads, holding power of anchor system

Ross Sea is driven towards and along the Victoria Land coast and that the major exit for surface waters is located in the north-west Ross Sea.

Bottom and subsurface currents are poorly understood. Current velocities were recorded at the J9 drill hole through the Ross Ice Shelf and reached 0.18 m/s, however, generally they were less than 0.10 m/s in the Ross Sea.[47,48] Current directions vary on the continental shelf, while on the continental slope flow is mainly north and west and parallel to topographic contours.[39]

Waves: The presence of ice generally reduces the height of waves. Waves tend to be bigger north of Ross Sea because of greater fetches and less ice. In the south wave heights are about 0.5 m and increase to 2 m further north, reaching 4 m during winter at 60°S.[49] Gale force storms over 12–24 hours and a fetch of several hundred km can rapidly create significant wave heights of 5 m.[39]

Ice conditions:[39] Antarctic sea ice is poorly understood and the lack of quantitative data is apparent. Existing records of observations are mainly from measurements in the austral summer and winter measurements are very scarce. In comparison to the Arctic less ridging and rubble fields develop in the Ross Sea. Ice distribution and concentration depends mainly on the Ross Sea polynya, which usually opens in late November along the front of the Ross Ice Shelf. In combination with the surface circulation, currents and winds this breakup pattern leads to a belt of pack ice to the north of the Ross Sea during January, with floe sizes of up to 10 km width. Due to the East Wind Drift, the Ross Sea Gyre and prevailing winds persistent pack ice conditions are created along Victoria Land, with drift rates between km/hr to 3.5 km/hr. Fast ice forms during most winters in a sheet 20–50 km wide off the coasts of Victoria Land and by early February most of the ice has broken out. Freeze-up starts in the south in early March reaching ice thicknesses on the average of 0.5 m by the end of April.

Icebergs: Initially icebergs move around the continent in the East Wind Drift and the Ross Sea Gyre, before moving north into the Antarctic Circumpolar Current. The drift pattern is usually very erratic and their exact drift pattern is difficult to predict. The western Ross Sea sometimes is referred to as an iceberg alley, with bergs smaller than 200 m long or wide being by size most frequent. About 3% of Antarctica's icebergs are larger than 1 km long or wide. The largest recorded iceberg of the Antarctic was sighted in the Ross Sea with a size twice of the state of Rhode Island.

Technical Considerations

Protection of platforms from ice and icebergs in deep water is the main problem associated with Antarctic petroleum development. The focal point of Antarctic petroleum development will be the design of offshore structures. No production platform for deepwater development suitable for severe sea ice and iceberg conditions is currently available. The understanding of the interaction between environmental forces, loading on the structure, water depth and geotechnical considerations is of utmost importance and is briefly outlined below.

Loading on a structure due to ice and iceberg impact is dependent on the active environmental driving forces. These forces – currents, wind, waves – are extremely strong in the Ross Sea (see table 2). The technical challenge for platform designs consists of how loads on a structure are transmitted to the sea bottom in deep water. Platform designs typically can be categorized into three concepts – artificial islands, bottom founded structures, and floating structures.[50] Artificial islands can resist almost any type of ice feature. Due to economics, however, they have an operational limit of 10 m (non-retained island) or 40 m (retained island e.g. caisson retained) water depth. Bottom-founded structures – typically gravity base structures, piled base structures and mixed base structures – are installed in intermediate water depths between 10 m and 100 m. Gravity base structures resist lateral loads solely by their large mass and the friction and shear developed at the base.[50] Pile-base structures are usually jacket structures or monopods, that develop shear resistance by means of piles. Mixed base structures combine shear resistance of a large base and lateral resistance of short piles to counter ice loads, and are commonly used for unfavorable soil conditions. Floating structures – moored or dynamically positioned steel or concrete structures and ice platforms – are used for deepwater developments. Stationkeeping is the main limitation. Ice platforms require stable, immobile sea ice conditions. Permanent flooding of the structure is required. to thicken the ice pad and to sustain its bearing capacity. Floating concrete and steel structures usually are equipped with a quick disconnect feature from a subsea well head when ice forces exceed design capabilities.

Three design configurations are possible: "brute strength" approach, "smart" system, and "evasive" design. A "brute strength" approach would be designed to withstand any environmental load at the given location. Such a solution appears to be impossible for Antarctic conditions. Technical knowledge on how to resist the mass and impact energy of a large iceberg (see table 2) does not exist at present and is currently not expected to be developed. A "smart" system could be built to recover from environmental impact (e.g. plastic deformation of materials). An "evasive" system would avoid the loading conditions of the severe environment, (e.g. a complete subsea completion system including wellheads, multiphase flow pipelines, subsea separation and production and subsea storage and transportation). Advances in the latter two approaches have been made over the last decade. The self-healing capacity of concrete to recover from cracks has been utilized by successfully installing concrete structures in the North Sea. Subsea completion is successfully being used in the Campos Basin in the Brazilian offshore area. Research in multiphase flow has advanced over the past decade and has been developed for the North Sea.[51] A combination of these two approaches could lead to a possible development scheme for Antarctica.

Global and local loads due to sea ice determine platform designs. Global load is the total force that an ice feature exerts on a structure. The pressure at the contact area between ice and structure is analogous to global pressure. Due to non-uniform distribution of this pressure a maximum pressure occurs within this area, which is analogous to local pressure. Pressure distribution depends on the geometry of the structure, its stiffness, and the behavior of the structure and the ice. High local pressure is due to grain size orientation and high confinement. The design of offshore structural elements is based on local pressures, while global pressures are used for foundation design.[52] Mechanical factors, like ice type, temperature, salinity, geometry of ice features, rate of loading, ice strength, elasticity, and structural factors (size and compliance) affect local pressures. However, the important factors are the strain rate, strength confinement and size of the contact area.[52]

Global loads are determined by calculating the limit stress load, limit momentum load and limit force load. Limit stress load is tied to an assumed infinite driving force on the ice. If an ice floe is brought to rest before the ice has failed across the full diameter of the structure then the ice load is less than assumed by the limit stress and is termed the limit momentum load. Iceberg impact forces are determined in this load calculation. If driving forces are limited directly then this limit to ice loads is called limit force load. Large diameter production structures are governed by this limit.[53] Measurements of sea ice properties that would allow calculation of these loads to a satisfying degree are currently not available for the Ross Sea.

Geotechnical conditions in the Ross Sea are very poorly understood. In deep water, severe ice environments can develop lateral forces on offshore structures that are one or more magnitudes larger than any previously dealt with offshore.[54] During winter, sustained and cyclic loadings are to be expected. Intense winter storms drive the ice canopy against the structure. As the ice breaks against the structure, cyclic loadings are superimposed on the overall loading. During summer, dynamic and cyclic loadings can be expected. These loadings develop as large ice features are driven by the winds, waves and currents into the structure. The general geotechnical implications affect the very complex patterns of strains that can be developed in foundation soils. Cyclic loadings, fatigue effects, and long-term creep effects can lead to degradations in soil stiffness and strength. Accumulated strains can lead to significant displacements of the structure.[54]

Future development needs to be concerned with frozen ground. The purpose of foundation bases and piles is to transfer deadweight of the structure to the foundation soil. In frost-susceptible soil, heaving may result in uneven uplift of piles or bases. Pile-jacking and base-heaving during freezing, as well as external freezeback pressure around piles and long-term creep and differential settlement are possible problems. Differential movement then redistributes the internal stress. This induces bending and shearing forces. Non-uniform lifting causes the structure load to be transmitted to soil subject to the greatest heave and differential settlement results in loads carried by soil with the least settlement, leading to overloading of some elements of the structure.[55] These heave forces act mainly on foundations. Adfreeze, the shear resistance at a soil/pile interface, and frictional loads act on lateral surfaces.[55]

Thermal disturbance of the permafrost composition might lead to thaw subsidence. Freezing and thawing gives rise to in-situ undrained strengths that might vary laterally as well as vertically. However, very little is known about the engineering properties of saline permafrost. The existence of brine pockets included in the soil column might lead to subsidence and loss of soil strength.[56] Relict permafrost can range from highly ice-bonded to non-bonded depending upon a sediment's composition. Susceptibility to thermal and mechanical degradation is possible. Once degraded, large changes take place in compression and strength characteristics.[57] Gas hydrates are solid, crystalline, ice-like combinations of water and gas (commonly methane) that form within the sediments. The existence of gas hydrates might lead to low soil strengths. For seafloor installations these conditions need to be known.

Overconsolidated soils are characterised by the fact that the present overburden stress is less than the maximum overburden stress the soil has been subjected to in the past. Sealevel changes throughout time contribute to overconsolidation, leading to unusual stress-strain characteristics of such soils. The settlement in a fine-grained overconsolidated soil needs to be estimated for structure selection, including the rates and time frame of consolidation. Freezing point depression and the amount of unfrozen water present at colder temperatures are important design considerations, when calculating rates for freezing and thawing and assessing the strength of frozen seabed.

The interaction between environmental forces, loading on the structure, existing water depth, and geotechnical considerations then leads to the conclusion that artificial islands and bottom-supported fixed structures are not feasible for economic reasons. It appears, that a floating structure in conjunction with subsea wellheads could be expected to be used for such development.

While in the case of a fixed structure almost all of the kinetic energy of the iceberg is absorbed in crushing, most of the energy in a floating structure is absorbed in elastic translation of the mooring system and a local elastic/plastic deformation at the contact point.[50] Therefore, platform designs require the evaluation of hull, riser and mooring systems. Elimination or reduction of iceberg impact can be achieved by either towing of the approaching iceberg, disconnecting the platform and moving it out of the way of the iceberg or resisting the impact of the iceberg, risking the loss of the platform. No design will be feasible for all iceberg sizes. However, technical design approaches are in development to increase the capacity of a platforms to withstand higher impacts. Avoidance of the impact all together is the most advantageous alternative. In the Arctic, small rounded icebergs sometimes are difficult to tow, since they cast off the towing sling or roll over. Quick disconnect features for riser systems and reentry systems do exist and could be utilized in the Antarctic.

The important parts of a platform that are vulnerable in this environment is the hull, mooring and anchoring system and the riser. The hull is the support structure for the topsides. Damage to the hull can easily lead to a loss of the entire platform. Ross Sea design will have to focus on special concerns regarding materials for such hull design. To provide for station keeping in such a harsh environment, a high strength mooring system is necessary. The great water depth in combination with the strong environmental forces of the Ross Sea will put high stress loads on the mooring lines in order to keep a floating structure within the maximum allowable offset design range. If the environmental forces exceed the systems' limits then the riser needs to be disconnected to prevent damage to riser and Blowout preventer. Downtime will increase rapidly with an inadequate mooring system. Such mooring lines still need to be developed. To provide holding power is the function of the anchoring system. Holding power of the anchor varies with soil conditions. Anchor deployment most likely will have to be in an omnidirectional geometry in order to respond to the dynamics of the local shifting environmental forces.

Logistics and Development

Logistics and development of offshore oil from the Ross Sea could look like the following scenario.

The closest usable marine base is at McMurdo Station – roughly 400 kilometers away. Christchurch, New Zealand is about 2500 km away. Only a very small percentage (3%) of Victoria Land can be used as staging area for personnel. Here will be an aviation base for the helicopters and landing strip for fixed wing airplanes coming out of Christchurch. A camp will be built onshore. This camp will also serve as the operational base for an oil spill center PIRO – the regional oil spill prevention center, suggested by the American Petroleum Institute[119] for several regional centers in the US.

Two drilling rigs could be employed to drill one well each per season. Supporting these two rigs is a major supply/logistics problem. A "wareship" concept (= warehouse ship) can be applied. This follows AMOCO's exploration scenario for the Navarin Basin.[58] Two vessels – one a converted bulk carrier and one a tanker will be the logistical center of the operation. The tanker carries all the diesel fuel for the rigs and vessels, fresh drilling water for the mud system and potable water for all the support vessels. The converted bulk carrier carries all of the bulk cement, barite, gel, casing, all drilling tools etc. Both vessels are self-propelled. These two ships then constitute the two main supply points close to the drilling rigs.

Five anchor-handling supply boats run supplies between these two major vessels and the rigs. One weekly run to Christchurch, New Zealand will be made with one of these supply boats to take back waste of the drilling operation, waste oil and garbage, and bring back fresh food supplies. Independence for the major supply materials from Christchurch is one of the main objectives of such a floating supply facility.

The key to the operation is the wareship, which needs to provide adequate capacity to handle a bulk system capable of transferring to the supply boats all of the bulk materials that are needed for the drilling operations, as well as a crane system that can operate in rough weather and a helideck.

Experience from the Arctic gives reason to believe that Antarctic exploration is feasible by utilizing two exploration platforms similar to the conical drilling unit "Kulluk",[59] which operated successfully in the Beaufort Sea. Such units will be towed to a drilling site in 500 m water depth. A total of three tugboats will be in operation, and in addition a salvage vessel. An icebreaker will be utilized for severe ice conditions. First a COST well will be drilled, followed by the drilling of four exploration wells. Eleven delineation wells will be drilled to confirm the existence of a discovery and to verify the extent of the field. The entire development is assumed to last 20 years, whereby prospecting will last 5 years, exploration, exploration drilling, and delineation 10 years, and final development 5 years.

For the development, a 400,000 Bopd scenario is assumed. Produced gas will be reinjected into the formation. The field life is assumed to be 30 years.

Eighty production wells will be drilled; two Tension Leg Platforms with forty production wells each. Each Tension Leg Platform is equipped for a 200,000 BOPD production capacity.[60] The wells will be drilled as subsea-completion wells. Individual well clusters will be drilled through a template. All wells will be sunk individually into glory holes. Through a central manifold for each template, production will be transported to a flexible riser system, that can be disconnected. A subsea storage facility for 12 million barrels will be built. A 300 km long buried subsea pipeline system will be built that links the individual templates with the production platform and the production platforms with the central storage facility. Five icebreaker shuttle tankers of 250,000 DWT each will transport the crude oil to an icefree port in New Zealand. Additional double-hull tankers will be used in times of no ice-occurrence or in dual utilization with an icebreaking tanker in the winter time, where the icebreaking tanker breaks a route for both of the tankers. A nine day round trip including loading and unloading is assumed. The icebreaking tankers, however, will transport the oil only to the icefree zone, from where regular tankers will transport the oil further to New Zealand. This means a four day return trip for these tankers. The tankers will be equipped with the SWOPS[61] system, which allows direct tapping of the storage facility by the tanker. In this way an offshore loading system is not necessary.

Exploration

Presently, investigations have been done for scientific purposes only. These investigations have included geophysical studies – gravity surveys, aeromagnetic surveys, multichannel seismic reflection surveys and drilling studies. A recent description of these studies and their relevance to petroleum resource potential is provided by Behrendt.[62] The type of work experience is compiled in the following summary (after source 62).

Geophysical: Since 1976 a total of 80,000 km multichannel seismic reflection profiles have been collected across Antarctic margins by 11 nations. In the Ross Sea the following 48 multichannel seismic data were collected:

year	collected data	organization
1980	6,100 km	BGR (Germany)
1982	1,500 km	IFP (France)
1983	N/A	JNOC (Japan)
1984	2,350 km	USGS (USA)
1985	48 km	
1987	4,300 km	

Drilling Studies

year	drillholes	organization
1973	4	Deep Sea Drilling Project
1979	1 (J9 – RISP)	National Science Foundation
1981	1 (MSSTS–1)	New Zealand
198?	1 (CIROS–1)	New Zealand

Choosing of an Exploration Drilling Vessel

The interaction of ice failure modes and structural shape and geometry of the platform determines the operating capabilities of a structure for a given environment. This interaction is described below. Ice deformation in front of a structure has a major influence on the ice load. The mode of ice deformation is determined by the structure shape and geometry and the geometry of the ice feature. A vertical structure generally will cause crushing as the failure mode of the ice, while bending will occur in the case of a sloping structure. Usually, iceloads caused by ice failing in bending are less than by crushing. A sloping structure, therefore, appears to be better suited for reduction of global ice forces. Croasdale[63] indicates three problems with such a sloping structure: (a) stationary ice may freeze onto the structure and this load can then be as high as the crushing load; (b) ice ride-up on the structure might occur (c) ice might fall back on the advancing ice and a rubble field will form, and the loads on the outside of the rubble will be the same as on a vertical structure. Frictional forces acting tangential to the contact area are important for designing sloping structures. The confinement that affects local loads is a function of the geometry increasing with the side slope angle.

In particular, a conical structure is appropriate. Gaida et al.[59] assess such a concept. An inverted truncated cone allows ice braking to occur in a simple fashion without having to rely on an active system such as a mooring turret. Such an omni-directional bow survival is high, even in the case of mooring failure. Also, such an inverted truncated cone increases the cargo carrying capacity and technology is advanced enough to build such a cone. Ice forces on structures are reduced by a sloping contact plane between structure and ice. With a decreasing angle of the contact plane a change in ice failure mode from crushing to bending results. Compressive strength of ice is considerably higher than its flexural strength, leading to the fact that ice forces decrease commensurately with decreasing angle from the horizontal. The fact that sloped contact planes reduce ice forces and the fact that ice forces against offshore structures can be exerted from all compass points, it suggests that a conical form is particularly favorable for such a structure.[64] Floating, moored, downward-breaking cones may take advantage of the structures compliance (i.e. ice induced pitching and heaving may support breaking of the ice at relatively low anchor line forces).[64]

The ice sheet failure for a downward breaking cone is accomplished by a combination of hydrostatic forces in submerging and moving the broken ice pieces around the cone and hydrodynamic forces that are developed as a result of relative movement between the ice and the water.[4] An inverted cone structure causes the ice to break downward and away from the vessel, protecting its drilling riser system and the mooring lines. Matsuishi and Ettema[66] investigated the forces and motions experienced by a floating, cable moored conical platform impacted by ice floes and mushy ice rubble. Their model resembled the drilling unit "Kulluk" on a 1:45 scale. The following results were obtained and summarized by Wessels and Kato:[64]

When undergoing continous impact with ice floes, the moored platform experienced its largest surge motion for relatively slow speeds; and the greatest fluctuation of mooring force was at slow speeds.

The maximum inertia force was approximately one third of the maximum mooring force, and the ratio of these forces decreased with decreasing speed of ice-floe impact.

For very slow impact of the floes, the restoring (surge and heave) forces experienced by the floating, moored platforms were almost equal to the restraining forces experienced by a fixed platform which was also tested for comparison. With increasing impact speed, the moored platform experienced smaller forces than did the fixed platform.

When impacted by ice rubble of mushy consistency, the moored test platform drifted horizontally and changed its trim, but did not oscillate as the field moved around it. The mooring force experienced by the moored test platform increased monotonically as a false bow of ice rubble developed around its leading perimeter. Once the false bow had reached an equilibrium size, ice loads remained more or less steady.

The drilling unit "Kulluk" successfully operated in the Beaufort Sea, demonstrating the advantages of a conical hull form in resisting ice forces. Pilkington et al.[67] reviewed "Kulluk's" performance after three years of operation. They found that "Kulluk" performed beyond expectations, especially in situations of thick ice in early summer (not model tested), and performed according to specifications in situations that were model tested.

Drilling

The drilling of one well is expected to last three months. Fully drilling, testing, and completing a well is very difficult to accomplish within one year.

Three functions need to be addressed in the drilling procedure: (a) create rock failure (b) transport cuttings to the surface (c) maintain the integrity of the wellbore.

Initiating failure of the rock requires that a stress needs to be generated by the drilling process that is greater than the strength of the rock. The geotechnical conditions of the Ross Sea do not indicate that this might be unusually problematical. Scientific drilling holes penetrated to depths of 700 m. A total of 6 successful holes were drilled. The experience gained from these shows that drilling can be successfully performed in the existing soil conditions (see above summary on drilling studies). Overconsolidated soils are common in other areas of the world and are successfully dealt with (e.g. Gulf of Mexico). Tills will slow down the drilling operation due to their unsorted nature, thus requiring frequent drilling breaks. Subsea permafrost should be drilled fairly rapidly in order to eliminate thaw and hole instability. Penetration of the subsea permafrost should be vertical, followed by directional drilling. After penetration of the permafrost and running of casing normal drilling operations can continue. Hydrate zones should be drilled with cooled mud at the equilibrium temperature of the hydrates and should be cased with high collapse-strength casing.[55]

Transporting cuttings to the surface is the function of the circulating fluid. It is the circulating fluid in combination with the warm temperature crude that contributes to thawing of the permafrost.

Thawing of the permafrost leads to potential loss of the integrity of the wellbore and collapse of the wellbore is possible. From a well completion standpoint, permafrost thawing and refreezing is the single most important factor for well design. Thawing and refreezing creates severe wellbore loading. Continued operations in permafrost require ultimate compressive strength; in particular if additional loads are induced by thaw subsidence and freezeback. Early strength for cementing purposes in permafrost is needed for short waiting times on cement. Freeze-thaw cycling is a concern mainly at shallow depths and at early times. Arctic cements must attain adequate bond strength with both thawed and frozen permafrost as a seal against pressure communication and unfrozen zone contamination. Thawed soil next to cement does not contract and pull away from the cement but is compressed inward due to lateral loads across the thaw front. Thaw subsidence loads are reduced when thaw is reduced. The use of gelled packer fluid in casing annuli is one means to limit thaw. The gel will prevent convective heat transfer. Refrigeration is used for absolute thaw prevention in the MacKenzie Delta. Refrigeration coils around double walled conductor casing, containing solid insulation is used to assure good bond between conductor casing and permafrost for well control purposes.[55]

A Bentonite–XC Polymer–KCL mud system with high yield point to plastic viscosity ratio was developed for MacKenzie delta operations. The KCl addition to the basic Bentonite–XC polymer mud serves to 1. increase the yield point, 2. inhibit mudstone and shale swelling, and 3. to depress the freezing point. High pH values are required to control corrosion associated with Cl ions. In order to prevent internal freezeback, waterbased drilling muds should be thoroughly displaced from internal casing annuli before placement of non-freezing packer fluids. Otherwise, damage to shut in wells can result from volume expansion upon freezing. Double walled insulated tubing/casing that significantly reduces permafrost thaw, and that could be run with normal handling procedures, should be used.[55]

Non-magnetic drill collar strings for directional drilling must shield the surveying compass from magnetized drill string, since the horizontal magnetic pull of the Earth decreases closer to the pole.

Well spacing: If well spacing is reduced for thaw subsidence, the beneficial effect of additional casing resistance from neighboring wells can more than offset the detrimental effect of increased thaw. Stiff casing, in addition to symmetry boundaries between wells, will resist the tendency for permafrost to form, and thus will magnify the freezeback problem. Thaw subsidence strains are small for close well spacings. As thaw increases around multiple wells, thaw zones eventually will coalesce.[55] Table 4 provides a summary of drilling problems, and was compiled after Goodman.[55]

Production

For the given environmental conditions, fixed platforms, such as a gravity base structure, would be a very logical choice. Such an approach was chosen for the Hibernia field in Nova Scotia. The existing environmental forces would best be withstood by such a platform. However, the water depth of the Ross Sea would make such a system extremely expensive. A floating production system, therefore, needs to be developed that, in conjunction with subsea completion wells, can counter these environmental forces.

A Tension Leg Platform (TLP) fulfills these requirements to a certain extent. However, no TLP has been developed that can operate in severely ice infested waters.

Demirbilek[68] summarized the main characteristics of a TLP. A TLP is a system resembling a semisubmersible, however, mooring a foundation is unique to this system. Tendons allow horizontal motion (surge, sway, and yaw) of the structure, but restrict vertical motions (heave, pitch and roll). Vertical columns and horizontal pontoons provide for the buoyancy of the structure. Excess buoyancy over the platform weight ensure permanent tension of the tendons in all weather and loading conditions.

The deck of a TLP is sensitive to payload increases, thus directly affecting its displacement requirements, which in turn will influence platform response characteristics. The basic mooring system consists of tendons and connectors. This is a permanent mooring system that holds the platform on station. Natural periods in the horizontal modes of motion are controlled by the pretension in the mooring system and the water depth. As water depth increases,

Table 4. Drilling Problems
(compiled after source 55)

Phenomenon	Characteristics	Comments
thaw subsidence	– as permafrost thaws, the soil can compact, and induce loads on the wellhead and wellbore system;	– need to prevent heat flow away from the wellbore
external freezeback	– thawed permafrost and waterbased; outside the casing will refreeze and generate inward radial loads around the wellbore; a complete refreeze is not required for pressures to reach significant levels	
gas hydrate decomposition	– if hydrates are decomposed in a confined system then high pressures can be the result	
permafrost lithology	– phase change contraction and thaw; consolidation in ice rich soil will generate vertical compression along the wellbore opposite these zones and tension above and below these zones; stiffness reduction and pore pressure reduction will induce alternating strains when lithology is layered; pressure load across the thaw front causes the relatively incompressible sands to expand vertically and compress silts	– determines nature of loads – induced by thaw subsidence
thaw discontinuities	– increased wellbore strains, pore pressure; reduction and/or stiffness reduction, can act across the horizontal thaw boundary of the properties discontinuity causing a vertical compressive sequence	– due to sudden variations in permafrost thermal
well spacing	– loss of vertical support from the thawed/ frozen interface; together with the increase in support from the concentration of casing this will affect the compressive and tensile strain response of the system	– can alter thaw subsidence and freezeback behavior; as well spacing decreases thaw – can coalesce between wells
sloughing/washout	– presents problems for cleaning, solids control, mud displacement positioning of insulating packer fluids, cement behind casing	– in near surface holes with excess ice hole instability – is expected and can be explained by melting of support provided by ice; in deeper soils hole enlargement is evident throughout permafrost and gravels are especially sensitive to washout
drilling through hydrates	– if mud becomes highly gasified this might lead to a well control problem	– the circulating fluid must be cooled to the hydrate equilibrium temperature; hydrate zones should be cased off with high-collapse-strength casing

frequencies of vertical motions enter the wave frequency range. This makes the mooring system one of the most problematic components of the design for deepwater. The foundation serves as the anchorage for the tendons and therefore keeps the platform in place. Foundation fixtures are secured to the seabed by either tension piles or gravity base structures. Installation of TLP foundations is especially challenging for deepwater development.

Main criteria for such an objective is to design a platform that consists of a hull that can withstand large iceloads and dynamic iceberg forces, a high capacity mooring system that can fulfill the function of station keeping in an adequate manner, and a riser system that is flexible and protected to operate in this environment.

Karsan et al.[60] describe a concrete TLP for sub-arctic conditions for the northern Barents Sea at 74°N latitude. This structure was designed to operate in a water depth of 500 m, to withstand a 500,000 tonne iceberg drifting at 0.6 m/s, and 1 m thick first year ice floes. The iceberg crushing strength was assumed to be 2 MPa for global force, and 4 MPa for local design. The platform concept consists of a five column TLP – four corner columns plus one central column for riser protection. Protecting of the TLP from ice impact is achieved by a fendering system, whereby impact energies are absorbed by plastic deformations of the cables, that span from the top of the pontoons to roughly 10 m above the water surface on the outside of the main columns. The column walls transfer loads acting on fenders to the overall TLP structural system. Pontoon weight leads to heavy internal stiffening against hydrostatic pressure. Columns, pontoons and bulkheads are made of lightweight aggregate concrete. Neutrally buoyant tethers are protected by locating the upper part of the tethers inside the fendering system.

Components of a concrete TLP are a ring shaped pontoon and four cylindrical columns piercing the sea surface and supporting the deck. Strength, durability and constructibility are the main objectives for using concrete TLPs. Increased strength has been achieved since the mid seventies.[69]

Advantages of concrete for floating offshore structures were described by Fjeld.[69] Concrete provides several advantages over a steel hull. Concrete is relatively insensitive to fatigue and corrosion, requires minimum maintenance requirements and exerts small loads on mooring systems and foundations. Concrete is impermeable and the problem of leakage can be disregarded. North Sea operations prove that no fatigue has yet occurred on 15 concrete structures. The impact of a 5,000 tons supply vessel ramming the platform at any speed can be withstood by a concrete platform. Selfhealing properties of concrete surface cracks reduce the possibility of corrosion of rebars. Concrete and cement react and deposit salts filling and closing the cracks. Lower stress levels in concrete hulls have been observed. Section forces and moments in floater hulls are typically 10–15% of the forces in fixed concrete platforms.[69] Experience from the North Sea has demonstrated an almost a maintenance free operation without material deterioration, rebar corrosion or other deterioration.

De Oliveira[70] described the advantages of concrete hulls for TLP's with respect to hydrodynamic forces. Concrete allows a lighter tether system because the increased hull dead load results in deeper draft and a lower center of gravity, and thus lower wave induced tether forces. A concrete cylindrical shell has a unique capacity to withstand large external water pressures. This deeper draft (TLP usually 50–70 m) can be achieved without significant additional cost. The response from vertical water particle motions is correspondingly low to deeper draft. Horizontal wave forces are resisted by mass inertia. As a wave force acts below the center of gravity, an overturning moment is created. This effect is reduced the lower the center of gravity occurs. This is the case of a concrete hull. This then has a significant importance for TLP tether forces.[69]

Such structures will not negatively affect the mooring system. Mooring forces do not increase drastically. Due to the increased draft of a heavier platform the tether forces are reduced relative to hull displacement. Increased hull displacement is achieved by increased draft.

Storage and Transportation

Storage needs to be done in seafloor storage facilities. Experience of such storage facilities is known from the Ekofisk storage tank in the North Sea and from the Persian Gulf. The storage tanks need to be constructed below the deepest recorded iceberg draft of 330 m.

Transportation of the crude will be by a combined pipeline/tanker configuration. Pipelines will transport the oil within the field and from the platform to the subsea storage facility. From this storage tank facility the crude oil then will be transported by tanker to New Zealand. The tanker size is assumed to be 250,000 DWT Average speed of such an icebreaking tanker is assumed to be 20 knots in open water but in ice conditions will decrease to 3–6 knots. Turnaround time at loading and unloading facilities is assumed to be 24 hours.

Buried, insulated marine pipelines are technically feasible. Designing a subsea-pipeline would include: prevention of permafrost degradation, protection from seabed ice scour, hydrodynamic stability requirements and cathodic protection requirements. No major technological advancements are required to install an Arctic marine pipeline.[71] For deepwater environments under ice conditions an icebreaker supply vessel and tug can be utilized.

ENVIRONMENTAL CONSIDERATIONS

Oil Spill Sources

The Minerals Management Service (MMS) requires, that intended offshore petroleum developments have to include an estimate of the number and size of oil spills that could occur as a result of the proposed activity, including a trajectory analysis for oil-spill movement. MMS data suggests that a significant oil spill resulting from loss of well control or other catastrophic events has a very low probability of occurrence. The MMS uses spill-rate constants to predict the number of spills most likely to occur from production operations, based on historical Outer Continental Shelf (OCS) operations. They are used to predict the number of spills likely to occur per billion barrels produced or transported.[72] For the Norton Basin[73] these numbers look like the following:

Spills per Billion Barrels of production

	≥ 1,000	≥100,000
Platforms	1.0	.036
Pipelines	1.6	.065
Tankers		
at sea	.9	.19
per port call	.2	.042

In the case of the Ross Sea these probabilities need to be multiplied by four to arrive at the respective figure of an oil spill probability for a 4 billion barrel oil field.

Prevention

Prevention measures as required for the Arctic can be applied to the Antarctic. MMS[72,74] established the following requirements for prevention of oil spills: Operators are required to use the best available and safest technology (BAST) for all activities when the MMS determines that 1. BAST is economically feasible 2. equipment failure would have a significant effect on safety, health, or the environment, unless their incremental benefits are clearly insufficient to justify the incremental costs of utilizing such technologies. The MMS considers that compliance with OCS orders, guidelines, and regulations constitutes BAST. The regulatory process provides for a continuous review and assessment of the equipment and technology proposed for use by lessees. These include requirements for the fitness of the drilling units, well casing and cementing, blowout-prevention equipment, mud programs, plugging and abandonment procedures, production-safety systems, well-completion and workover, pollution-control equipment, platforms and structures, pipelines and hydrocarbon handling and-storage equipment.

Clean-up Technology

MMS usually considers mitigation of major oil spills possible through the tenth day of a spill, after which the oil would be too dispersed for effective recovery. MMS considers a winter spill, that would freeze into the ice and remain relatively intact until it melted out as a fresh, early summer spill.[74] Mechanical cleanup is considered much more effective on low- or medium-viscosity oils than on high-viscosity oils. Viscosity, however, can change rapidly (4 hours). from a low viscosity to an emulsion state (e.g. Prudhoe Bay crude) given that broken ice conditions exist. In the absence of sea ice this process might take two days.

Dispersants can be used. However, the effectiveness decreases far more rapidly than mechanical cleanup possibilities as oil weathers and becomes more viscous. Kirstein and Redding[75] calculated an 8 hour period for effectiveness of dispersants for a summer spill of 22,000 bbl Prudhoe Bay crude. In the presence of sea ice the formation of mousse would shorten this interval. The perceived toxicity of oil dispersant mixtures makes it doubtful that they would be allowed to be applied in Antarctic waters. Burning of the oil is a mitigation measure. However, immediate action is required to achieve an effective rate of 50 to 60%.[74]

The effectiveness of mechanical and burning decreases with increasing sea state. However, effectiveness of dispersants and natural dispersion increases.

Planning an effective surface response with mechanical equipment to spills in pack ice would require that an icebreaker (or icebreaking-supply ship) be stationed locally in both winter and summer as a dedicated oil-recovery vessel.[74] The ice-managing supply vessels during the exploration phase could be used for this function. Burning experiments in broken ice have given promising results with fresh oil, but results have been variable and less promising with weathered oil and emulsions. Field tests in a mud pit at Prudhoe Bay were able to burn 55% to 85% of fresh Prudhoe Bay crude, but sparged crude with a flash point of over 30°F could not be ignited. Tests for fresh or sparged crude had burn efficiencies of 85% to 95% at 22% to 34% ice cover and burn efficiencies of 58% to 79% at 78% to 85% ice cover. Burn efficiencies of two tests for oil-in-water emulsions were only 10% to 52% at 78% to 84% ice cover.[74] Due to fast emulsification, a slick would have to be set on fire very soon after spillage in order to obtain a high burn efficiency. Burning of oil might be more difficult during ice freezeup, since wave action mixes the oil downward into the grease ice. Oil and ice would have to be recovered and the oil separated from ice before burning; there would be only a limited capability for in situ burning.[74]

The fact remains that the capability to clean up large oil spills floating amongst moving ice is generally not good, particularly if the oil is thin and weathered.

Oil and Sea Ice

Wind and ice motion would concentrate oil in leads. Leads that form afterwards are probably less affected by a spill.[39]

Sub-sea-ice oil might become trapped due to further freezing of a new ice layer underneath, thus forming pockets that after ice break-up then might be released as fresh oil after ice break-up. Thus, initially a spread is limited, but it could travel long distances after break-up, since sub sea-ice treatment seems very difficult.

In the Ross Sea, oil spilled against ice cliffs, fast ice and shorelines probably could be physically removed relatively rapidly due to wave action.[39] Ice could serve as a protective mechanism for rocky coasts and beaches, limiting the retention of oil on coastlines.[39]

Because of the great extent of the pack ice it would require an enormous under-ice hydrocarbon spill or blowout to affect more than a local part of the Ross Sea, although break-up, currents and wind would move the oiled ice from the spill site.[39]

When oil is trapped into ice, weathering processes such as evaporation, dissolution, biodegradation and dispersion are significantly retarded or even stopped. Hence, when released after thawing, the oil will be relatively fresh. Emulsions formed in the presence of developing ice have significantly higher viscosities than those formed under ice-free conditions.[76] Payne et al.[77] observed that extremely rapid formation of stable water-in-oil emulsions occurred under ice-forming conditions, which they attributed to low water temperatures and to the microscale turbulence created by the grinding of the grease ice crystals, which injected small water droplets into the viscous oil. On thawing, these emulsions were sufficiently dense to cause them to reside immediately below the grease ice. With continued agitation and gelting, the emulsions eventually surfaced into patches of open water between the individual ice floes.

Wilson and Mackay[78] concluded that significant quantities of oil may be entrained within a developing ice field under freezing conditions. The extent of oil incorporation was enhanced by: a high oil density and/or viscosity occurring naturally or induced by weathering; the presence of sufficient turbulence to disperse the surface oil and to induce mixing within the ice field; the formation of emulsions of seewater and/or ice in oil; the formation of small oil droplets; and the formation of coalesced ice particles measuring about 5 mm in diameter.

These authors also concluded that the densest oils would be released relatively slowly from a frozen pancake ice during thawing. Once the oil begins to collect on the surface of the ice, solar radiation will tend to hasten the thawing process.

Weathering Processes

Table 5 summarizes the possible fate that oil will undergo by the individual physical, chemical and biological hydrocarbon degradation processes. These processes determine the long-term effects of a spill.

Effect of an Oil Spill

Beaches would be sensitive to oil spills during the austral summer when protection by shoreline ice is reduced. Keys[39] suggested that coarse porous beach materials could retain much oil through penetration and burial. It is estimated that the beaches at Cape Bird could retain well over six million barrels, some of which would be slowly released by erosion and washing during the following decades. Thus of the coastal types recognized, the environmental risk is likely to be greatest for beaches[39] McWhinnie[79] and Zumberge[80] consider that a large oil spill in the Ross Sea, involving up to hundreds of thousands of barrels, would result in the loss of no more than 0.0001 percent of the total

Table 5. Weathering Processes

Process	Effect	Source
spreading and drift	– strong circulation will spread and dilute the contaminants, possibly raising background levels	121
	– onshore winds, currents and the Ekman Drift could transport suspended matter onshore; if this effect is stronger than the natural removal rate then beaches and shallow benthic communities are at risk	121,122
	– strong horizontal, vertical and lack of stability of the water column should promote dilution of any pollutants to acceptable levels and spread them northwards away from the continent	121,80
	– most oils probably float for a considerable time in Ross Sea	39
dispersion	– dispersion to the north can be expected in one year, which would be in line with first year ice that is dispersed to the north each year	80
	– high wave energy enhances the formation of small droplets and this small size allows the droplets to disperse rapidly resulting in ever-decreasing concentrations; however, a more viscous oil will disperse slower	76
emulsification	– enhanced by low temperatures and high energy conditions	76
	– slicks of highly viscous oils/emulsions are relatively persistent, due to their high cohesiveness they will break up very slowly and eventually will form tarballs	123
biodegredation	– information on naturally occurring hydrocabon is conflicting around a drillsite on Ross Island where diesel fuel was used as a drilling fluid the indigenous populations of bacteria, fungi and algae were completely eliminated and replaced by a few entirely different species of what were believed to be hydrocarbon decomposing bacteria	124
	– evidence for hydrocarbon-degrading bacteria at the New Harbor and Commonwealth Glacier drill sites was shown by using various growth media including Bushnell-Haas hydrocarbon agar	125
	– no evidence that microbes growing in oiled soil from Scott Base and McMurdo Station could have biochemically decomposed the oil at those sites	126
	– these bacteria that caused biodegradation could have been merely growing on contaminants or in the presence of contaminants or in the presence of oil	39

Process	Effect	Source
	– biodegradation of oil occurs in the Arctic but at very slow rates	127
	– temperatures in the Ross Sea are colder than in the Arctic, thus oil biodegradation processes might be even slower there (provided hydrocarbon-degrading microbes do exist there)	39
	– temperature was not a limiting factor for biodegradation in the Strait of Megallan oilspill	128
	– dodecane oxidation rates were 0.7g/l/day in Port Valdez, 0.5g/l/day in the Chuckchi Sea, and 0.001 g/l/day in the Arctic Ocean	129
	– experimental results of biodegradation were found only after one year in the case of Prudhoe Bay crude in Beaufort Sea sediments	130
	– average degradation rates for n-hexadecane in experiments on Baffin Bay Island varied from 9.5 to 43.8 microgram/m3/day for waters with the highest rates occurring in early August	131
evaporation	– one of the primary removal mechanisms on Palmer Station, Antarctica	82,83,84,85
photooxidation	– very effective in the summer time, while in winter time the effect will be drastically reduced	80
dissolution	– volatile low molecular weight compounds, which are relatively toxic dissolve in seawater	132
	– a large proportion of these dissolved hydrocarbons would evaporate on the order of two magnitudes greater than solution	
absorption	– in general, sediments with high clay and organic contents are likely to concentrate and absorb petroleum hydrocarbons; such sediments are not widespread inshore but are common in deeper waters of the Ross Sea	39

Antarctic krill biomass. Near shore spawning areas should be regarded as vulnerable.

Experience elsewhere, including the Arctic, suggests that most non-furry marine mammals generally are not particularly sensitive to oil. Therefore, fur seals are the only mammals likely to be particularly sensitive to the mechanical effects of oil pollution in the Antarctic.[81] Their range does not include the Ross Sea.

Penguins and birds will be endangered, since death is caused by the loss of water proofing and buoyancy resulting from oiling of feathers. In the case of penguins, because of their vast population, estimated to be on the order of 120 million individuals, there would appear to be adequate numbers to replenish those lost, even if the casualties were numbered in the tenth of thousands.[80]

The same is true to a lesser extent for seals. In particular, oil accumulations in nearshore areas would effect adult seals and newborn seals. In the open sea only transient effects are expected.[80]

The shallow benthos is a vulnerable habitat particularly in enclosed waters with fined-grained sediments, where bioaccumulation in benthic organisms is possible.

The Scientific Committee on Antarctic Research concluded that the impact of spillage from a fully laden oil tanker would be very localized because of the immense diluting properties of the Southern Ocean with an area of 36 million km^2 south of the Antarctic convergence. This area is compared to an area 1.5 km^2 for the Gulf of Mexico and 575,000 km^2 for the North Sea. The report emphasizes that while local impacts may be severe in terms of mortality, they would not have a significant impact on the standing stocks of marine organisms. If an oil spill covering 300 km^2 were to occur in an area of dense swarms of krill, about 15,000 metric tons of krill would be endangered. Assuming complete mortality, the loss as a proportion of the total krill population would be insignificant. Moreover, annual reproduction, estimated to be on the order of 75 to 100 million tons, would be more than adequate to replenish this loss.[80]

Zumberge states that all that can be said is that a spill of 2.5 million barrels of oil in Antarctic waters could have an environmental impact ranging from negligible to serious.[80]

Experience from an Antarctic Oil Spiill

On January 28, 1989, the Argentinean ship Bahia Paraiso ran around near Anvers Island on the Antarctic Peninsula spilling more than 150,000 gallons of petroleum, primarily Diesel Fuel Arctic (DFA). Kennicutt et al.[82,83,84,85] reported the results of the investigations.

Microbiological studies suggest that indigenous densities of hydrocarbon-oxidizing bacteria and degradation rates are extremely low compared with temperate climates. Microbial hydrocarbon oxidation was measured by monitoring the rate of 14C–CO^2 evolution from incubations of seawater, plankton concentrates and sediments enhanced with radiolabled 14C-hexadecane (n–C16).

Microbial hydrocarbon oxidation occurred in sediments at very low rates (<100 g/C16/cm³/yr). The effect on benthic fishes appeared to be negligible. The effect on area mammals and seals also appeared to be negligible. The impact of the spill was limited due to factors like volatility of the released product, the volume of material released, the variable and severe weather conditions, the common occurrence of offshore winds and currents, and the lack of low energy intertidal areas for the fuel to accumulate. The primary removal mechanisms were evaporation, dilution, and transport from the area, with only minor effects from microbial oxidation, photo-oxidation, and biological uptake. Longer term sublethal effects appeared to be limited to limpets and bird population dynamics.

ECONOMIC EVALUATION

General Economic Setting

The economics of Antarctic minerals and their exploitation are a function of the differential in the costs of discovery, production, transportation, marketing, organization, and money.[86] Parameters that influence these differentials are mainly environmental conditions, distances between supply sources and markets, non-existing infrastructure and labor supply, special costs for discovery and evaluation, special managerial costs for control and supervision, and the high level of risk.

Capital and operating costs will be distinctly higher in this environment than in other areas. Potter[87] determined that probably the highest barrier to commercial operations on Antarctic shores will be the large overhead pertaining to icebreakers, equipment and labor in establishing a beachhead or preparing and maintaining airfields.

In order to forecast at what time in the future Antarctic minerals will be profitable one should determine with the highest possible level of accuracy the physical nature of the resource base, production costs, changes in world supply, demand, real price changes over time, trends in technology development, and new discoveries, and a multitude of other factors such as international mineral rights.

Risks of Development in Antarctica

The major risks in developing an oil field in Antarctica with the characteristics hypothesized here are obtaining a sufficient and timely supply of capital and obtaining an adequate return on that investment in the future market situation.

Cameron and Polio[88] identified five risks that commercial lenders are most aware of: equity risk, completion risk, operating risk, market risk, and supply risk. Equity risk – the possibility that the project might not generate satisfactory returns, and the completion risk – the sponsor's capabilities to meet the stipulations of the financing agreement verified by meeting of completion tests, are assumed by the sponsor. Operating risk – mainly technical factors that affect operations, market risk – that the actual market price is less than the cash flow analysis assumed, and the supply risk – the increasing cost for material input, are assumed by the commercial lender once completion tests are satisfactorily fulfilled.

However, two basic elements are closely evaluated by the lender: firstly, the operator's ability to contain costs within the specifications of the loan contract and secondly, the likely price trends throughout the term of the loan. Two independent ways are used by lenders to evaluate a project: evaluation in terms of unchanged real prices, meaning corrected for assumed rates of world inflation, and in current prices. Current prices are used to evaluate the deal, while constant prices are used to evaluate the mine. Predicted price spikes by the sponsor are frequently not fully supported by commercial lenders. Instead a more conservative approach is followed by commercial money institutions.

For VLB petroleum development to be economical, projected price spikes accompanied by a tightening of the supply side of the existing market will be necessary. Under these premises it is obvious that an extremely high risk investment results. In response, the operator would most likely have to finance the project by 100% equity. Even though a promising market development for petroleum might be assumed it is obvious that no forecasting – especially for 50 years into the future under these specific conditions – can be expected to be very reliable. Short term fluctuations of supply and demand and prices will always occur. Table 6 shows the cost and production structure of the project.

Evaluation Methods

Petroleum development projects are commonly evaluated by determining a minimum rate of return that the project should yield or by calculating the earliest possible payback period. Discounted cash flow methods are techniques used to arrive at this objective.

Usual methods for determining the profitability of a project are achieved by applying discounted cash flow include: net present value, internal rate of return, and payback period. These different methods should be analyzed together since various aspects of the investment are measured by different methods.

The discount rate can be determined as the cost of capital in the sense that it can be used to determine what the use of that capital in other investments might earn. Therefore, the discount rate can be set equal to the minimum acceptable rate of return of such an investment.

Cash flow is calculated by totalling the revenue of a specific time period, usually one year, and subtracting all the expenditures of that time period from it, including taxes and interest payments. Depreciation and amortization allowances reduce the taxable income but add to the cash flow after taxes thus increasing the total cash flow.

To calculate the net present value (NPV) each positive and negative cash flow is discounted to present value, using the appropriate discount rate. The sum of all present values of these cash flows gives the net present value of the

Table 6. Cost and Production Structure of the Hypothetical Oil Field[116]
(compiled after source 116 unless otherwise indicated; figures in paranthesis indicate current dollars)

CAPITAL COSTS

#	Item	Costs(US$)/ unit
2	Exploration structure	160,000,000 (1985)
1	Deck Barge (10,000 tonnes DWT)	5,200,000 (1985)
1	Deck Barge (15,000 tonnes DWT)	7,800,000 (1985)
1	Deck Barge (20,000 tonnes DWT)	10,300,000 (1985)
1	Deck Barge (25,000 tonnes DWT)	13,300,000 (1985)
1	Icebreaking Deck Barge (8,900 DWT)	7,100,000 (1985)
1	Tugboat (4,500 BHP)	6,400,000 (1985)
1	Tugboat (6,000 BHP)	7,400,000 (1985)
1	Tugboat (7,200 BHP)	8,300,000 (1985)
1	Salvage Vessel (2,250 BHP)	2,800,000 (1985)
1	Icebreaker (Class 6)	100,000,000 (1982)
1	camp (onshore)	50,000,000 (1985)
	permanent camp	100,000,000 (1985)
2	Production structure	300,000,000 (1990)[60]
5	Class 8 icebreaker tanker	360,000,000 (1982)
1	Storage facility	250,000,000 (1988)[117]
1	pipeline (300 km)	700,000,000 (1982)
1	Piro oil spill center	100,000,000[119]
	working capital	15%

OPERATING COSTS

Unit	Item	Costs(US$)
	Exploration activity	
1	Exploration structure[a]	2,500,000/yr (1985)
1	Deck Barge (10,000 tonnes DWT)	250,000/yr (1985)
1	Deck Barg (15,000 tonnes DWT)	320,000/yr (1985)
1	Deck Barge (20,000 tonnes DWT)	400,000/yr (1985)
1	Deck Barge (25,000 tonnes DWT)	460,000/yr (1985)
1	Icebreaking Deck Barge (8,900 DWT)	250,000/yr (1985)
1	Tugboat (4,500 BHP)	1,060,000/yr (1985)
1	Tugboat (6,000 BHP)	1,100,000/yr (1985)
1	Tugboat (7,200 BHP)	1,140,000/yr (1985)
1	Salvage Vessel (2,250 BHP)	560,000/yr (1985)
1	Icebreaker (Class 6)	15,000,000/yr (1982)
1	camp (onshore)	10,000,000/yr (1985)
	permanent camp	18,000,000/yr (1985)

Unit	Item	Costs(US$)
1	Production structure	2,500,000/yr
1	Class 8 icebreaker tanker (250,000 DWT)	57,000,000/yr (1982)
1	Storage facility	1,000,000/yr*
1	pipeline	25,000,000/yr
1	Piro oil spill center	1,000,000/yr*

* assumed by author

EXPLORATION COSTS PER WELL

(1 COST well, 4 exploration wells, 11 delineation wells)
Depth:10,000ft
Operating time: 10 years except exploration structure: 8 years

	$ US
Capital Cost/Well	
Exploration Structure	20,000,000
Consumables and Manpower	7,200,000
Support Vessels	10,500,000
Camp	3,125,000
Annual Operating Cost/Well	
Exploration Structure	2,500,000
Support Vessels	13,000,000
Camp	6,250,000
Total Cost/Exploration Well	63,000,000
Exploration Cost (excl. seismic)	1,008,000,000
Seismic costs/km	300[118]

PRODUCTION COST/WELL

(80 Producing Wells)
Depth: 10,000ft
Operating time: 30 years

	$ US
Capital Cost	
Production platform	7,500,000
Consumables and Manpower	6,500,000
Gloryhole	250,000
Flowlines/Pipelines	8,750,000
Icebreaker Tanker	22,500,000
Oilspill Center	1,250,000
Storage Facility	3,125,000
Camp	1,250,000

	$ US
Operating Cost	
Production Platform	2,000,000
Flowline/Pipeline	10,000,000
Icebreaker Tanker	107,000,000
Storage	1,000,000
Oil Spill Center	1,000,000
Camp	6,750,000
Working Capital/Well	26,000,000
Total Cost/Producing Well	205,000,000
Development Cost	16,400,000,000
Total Cost (Exploration and Development	17,408,000,000

Assumption: If Antarctic development is 1.5 times[120] as expensive as the Arctic then
Final Total Cost   26,112,000,000

Production (total):	4,320,000,000 barrels
Production/Well	54,000,000 barrels
Production/Well/Year	1,800,000 barrels
Production/Well/Day	4,630 barrels

Production cost/ barrel	$6.5

investment opportunity. The net present value is the indicator of the net yield of an investment, by determining the maximum amount that can be paid for an investment and still receive the desired minimum rate of return in its present value. Thus, the obtainable objective is a positive net present value.

The discounted cash flow rate of return (DCFROR) or internal rate of return (IRR) is the average percentage return that an investment opportunity is expected to yield over its life. This is evaluated by equating the present value of the positive cash flows with the present value of the negative cash flows, and determining that discount rate that produces the equality. The minimum requirement is a rate of return greater than the discount rate. Of course, the cash flows have to be adjusted for inflation.

The payback period is the number of years required to recover the original investment plus working capital at a zero percent interest rate.

Scenarios

This study focuses on NPV, IRR, and the payback period. The original base case and five alternative scenarios form a sensitivity analysis. The base case adheres strictly to average figures and production rates common in offshore operations. The base case is quite uneconomical. Those parameters that are most speculative were then changed and incorporated into the scenario to determine what conditions need to be present to make the project economical. These variations represent the following combinations of factors that influence the final evaluation:

Scenario A determines the influence of the initial capital outlay as a major cost factor for cash flow calculations. Since a tremendous initial investment is necessary in order to start operations, it is of utmost importance to determine the sensitivity of the capital costs on the final outcome of the venture.

Scenario B investigates how much the operational costs must decrease for a profitable operation to be possible when holding revenues are fixed. Since operational costs in this harsh environment do have a major impact on the profit-loss situation, it is important to determine their sensitivity.

Scenario C determines how high the required price must go for a profitable operation throughout the project's lifetime when operational costs are assumed to be fixed. Since price developments are critical for revenue generation, it is important to find lower price limits necessary to keep the operation profitable.

Scenario D determines how high the required price must go for a profitable operation throughout a project's lifetime, when an oil spill costing $2.5 billion is assumed to occur in year.[9] Environmental concern and therefore liability issues are of great importance for Antarctic development.

Scenario E is similar to scenario D, but assumes an oilspill in year 19 of the operation.

Assumptions

In order to evaluate this project certain assumptions have been made. Assumptions were based on similar operations, past experience, current trends, and the state of technology. Assumptions made are as follows:

(a)   Mineral and entry rights will be granted and the land will be leased;
(b)   The life of the project will be 50 years from the time of initial investment;
(c)   Financing of the project will be strictly from the company's own equity capital;
(d)   Salvage value for all plant and equipment will be zero. No additional equipment will be needed and any upgrading or repair will be funded out of working capital;
(e)   Land is to be leased and there will be no salvage value of land;
(f)   Inflation will be figured at a constant rate of 5%;
(g)   There will be no escalation of oil prices;
(h)   Any changes in the tax system will have insignificant effects on the project;
(i)   Taxation will be calculated at 30%;
(j)   Production will decrease at 5%/year starting with 144,000,000 barrels per year;
(k)   Labor, both skilled and unskilled will be available;
(l)   Working capital is figured at 15% of total operational costs;
(m)   Depreciation will be to straight line;
(n)   A real discount rate of 10% is assumed;
(o)   Expenditures for operational costs are capitalized in the twenty years of the development;
(p)   All capital investment will be done in years zero through nineteen;
(q)   No depletion is assumed, since this is a regulation only allowed in North America. Since the tax code for Antarctica is unknown this allowance is not incorporated in this cash flow analysis.
(r)   Adjustment for inflation of the individual cost components to be expressed in constant dollars for one particular year in the cost structure was not considered relevant, since many cost estimates of individual items are very speculative.
       A further distortion of the figures would be the result, since many items (e.g. ice breaking shuttle tankers) have not been developed yet.
(s)   All costs for individual development stages were added and then equally distributed over the time that the specific development phase would last

Results

Results of the individual scenarios are summarized below.

BASE CASE   **Initial investment 7.6 billion US dollars**
                   NPV @ 15% Real Discount Rate   − $ 705,217,058
                   NPV @ 10% Real Discount Rate   − $1,117,480,922
                   NPV @ 5% Real Discount Rate   − $1,437,150,004
                   NPV @ 0% Real Discount Rate $ 706,238,472
                   IRR   <0%
                   Payback Period   ________________

SCENARIO A   **Maximum initial investment that guarantees a profitable outcome (or a decrease in the base case by 70%)**
NPV @ 10% Real Discount Rate $ 72,136,374
IRR   11.30%
Payback Period   25 years

SCENARIO B   **A decrease of operating costs by 100%**
NPV @ 10% Real Discount Rate   – $927,621,291
IRR   <0%
Payback Period   ______________

SCENARIO C   **An increase in commodity prices by a factor of 2.5 to $50/bbl, while keeping all other factors constant**
NPV @ 10% Real Discount Rate $ 844,762
IRR   10%
Payback Period   25 years

SCENARIO D   **A required oil price of $76/bbl that guarantees a profitable outcome in case of an oil spill of $2.5 billion in year 9**
NPV @ 10% Real Discount Rate $ 6,202,131
IRR   10.02%
Payback Period   25 years

SCENARIO E   **A required oil price of $73/bbl that guarantees a profitable outcome in case of an oil spill of $2.5 billion in year 19**
NPV @ 10% Real Discount Rate $ 36,971,742
IRR   10.15%
Payback Period   26 years

The result of the base case scenario makes it immediately obvious, that no profitable offshore petroleum development can be conducted in the Antarctic under current price and technical conditions. All of the evaluation methods led to negative results. Commonly, an increasing discount rate will result in a lower NPV. However, the base case scenario illustrated a reverse picture: increasing discount rates resulted in a more favorable NPV starting with a discount rate of 15%. An abnormal NPV profile at discount rates of 0%, 5%, 10%, 15%, and 30% is then the result. This is due to the necessarily very high initial investment in the beginning of the operation, which then is followed by very small positive cash flows throughout the remainder of the lifetime. The choice of the discount rate on the present value of cash flows, clearly is much more important in the first few years than in later years. In comparison to the tremendously high negative cash flows in the beginning of the project, the discounted positive cash flows become fairly small. It has to be pointed out, that, even though increasing discount rates improve the NPV drastically, the final result will nevertheless remain negative. Therefore, it becomes obvious that the IRR is less than zero and no payback period is indicated.

A break-even analysis of the base case scenario was performed, to determine the necessary production. It showed, that a total production of crude oil of at least 360,000,000 bbl per year, declining at 5% per year, is required to break even. The break-even analysis shows, that the assumptions made in the base case will not provide for a profitable development unless a drastic increase of the production rate occurs.

In *Scenario A* it was determined to what extent the initial investment has to decrease in order to make this project an economically viable operation.

The results indicate, that at a real discount rate of 10% the initial investment has to decrease by 70% before this operation becomes profitable. This fact by itself indicates already the impossibility of such a venture. An NPV of $72,136,374 was calculated and the IRR of 11.30% indicates that this operation would be profitable and possibly preferable to other investment opportunities at that discount rate. A fairly long payback period of 25 years emphasizes the riskiness of such operation. It has to be emphasized that a decrease in capital costs of 70% is very unlikely to occur.

In *Scenario B* it was determined to what extent the operating costs would have to decrease before a profitable operation would be possible.

Decreasing the operating costs by 100% will not lead to a positive NPV, and the IRR is less than zero, even when operating costs are 0% of the base case. The negative NPV of $927,621,291 is not very different from the base case NPV of negative $1,117,480,922. This indicates the tremendous effect of the operating costs.

This indicates that under the given assumptions, investment in this operation could not lead to a positive outcome. This result shows the tremendous impact of the operating costs.

In *Scenario C* price of oil was increased by a factor of 2.5 to $50/bbl, leading to a higher gross revenue, while all other factors remained constant. At a 10% discount rate an NPV of $844,762 was calculated. This shows, that the high initial investment can be compensated for by an increased market value of petroleum. The IRR of 10% indicates a profitable operation under these conditions. This is combined with a fairly long payback period of 25 years.

Overall, it does not seem unreasonable that commodity prices could rise to these levels of this scenario. Crude oil prices were more than tripled by OPEC in 1973, exceeding the factor of 2.5 assumed in this scenario. The sensitive market conditions for petroleum were demonstrated by the 1991 Gulf war, when oil prices rose rapidly from the mid 20s to the mid 30s $/bbl within two weeks of diplomatic negotiations, but dropped within two days below $19/bbl after the invasion of Kuwait by allied troops.[89] This scenario is not out of the range of possible developments in the commodity market. Market developments with upwards moving commodity prices do provide a possibility for such a venture. Clearly, even if prices rise according to this scenario the profit would still remain marginal considering the high risk that has to be taken.

In *Scenario D* the effect on the project by an oil spill in year 9, costing $2.5 billion, was evaluated. The price of oil had to increase to $76/bbl in order to allow for feasibility of this project. This means that in order to be prepared for such environmental disaster crude oil prices have to quadruple. The calculated NPV was $6,202,131. The IRR was 10.02%. The payback period was 25 years.

In *Scenario E* evaluated scenario D under the condition that the oil spill would happen in year 19. The required oil price in order to guarantee a viable operation was calculated to be $73. This is due to the effect of the additional 10 years of discounting the occurrence. The calculate NPV was $36,971,742. The IRR was 10.15%. The payback period was 26 years.

GEOPOLITICAL CONSIDERATIONS

The Antarctic Treaty does not cover mineral resources. However, since 1970 at the Sixth Consultative Meeting in Tokyo the issue became of importance resulting in several recommendations on how to deal with the subject. In the Eleventh Consultative Meeting in 1981, it was recognized, that a specific minerals regime was a matter of urgency and recommendation XI–1 was adopted, setting guidelines for a future minerals regime (see below).

De facto it seems that the countries involved agreed on a ban on commercial activities. Clearly, however, it has to be recognized that the individual parties involved have different views on how to set up a mineral regime. These major interest groups are the treaty powers, the environmentalists and the Third World Countries.

The current interest in Antarctic minerals is not of an economical nature. Rather, it is much more a concern to avoid falling behind and loosing future access to potentially important resources. This is clearly realized by all the interested parties, leading to the recognition of the urgency of a mineral regime. Beeby[90] gave as the major reason for doing the job quickly, that if the issue remains unsolved it may present a threat to the Antarctic Treaty and the Antarctic Treaty System. Once the arrangements have been set up it will make it immensely difficult for a non-participating country to join a group of experienced countries, once the race for resources reaches the exploitation phase. It is generally feared by the Third World Countries that a mineral regime set up by the Treaty powers will lead to an advantage to the industrialized countries. This is also being nourished by quoting R.Tucker Scully, the chief US negotiator, by saying, that 'there will be no milking of Antarctic royalties and there will be no free lunch'.[91] However, the Treaty powers themselves are in disagreement on how to treat the sovereignty question. Due to the fact that no prohibition in the Antarctic Treaty is in existence to prevent commercial activities by companies of countries that claim territories in the Antarctic, activities of such kind are at least possible. It becomes even more complicated when taking into account that companies of countries that are members of the European Community would be allowed to operate in the British and French-claimed sectors as well. On the other hand, recommendation IX urged their nationals and other States to refrain from all exploration and exploitation of Antarctic mineral resources while making progress towards adoption of a minerals regime.[92]

This concept of 'voluntary restraint' is arguable, as geological groundwork is scientific research with a clear exploration and exploitation character in later stages and thus "exploration for mineral resources in Antarctica has already begun".[93]

Third World nations clearly take another view of the issue. This group favors the idea of internationalizing Antarctica and to treat it as a 'commons' similar to the seabed or space.[94] In general, the view of the UN Declaration of Principles is being followed by applying the principle of the common heritage of mankind "for the benefit of mankind as a whole, taking into particular consideration the interests and needs of the developing countries".[94] Similarities between the seabed and Antarctica are accepted as both deal with areas beyond recognized national jurisdiction. Mitchell and Tinker[94] pointed out that perhaps a precedent has been set by the Law of the Sea negotiations for internationalizing Antarctica and setting up a machinery for administering it. An argument for Third World Nations was made by the Malaysian and Sri Lankan Ambassadors to the UN General Assembly, where the common good of world resources was emphasized rather than the benefit of a few.

Environmental groups perceive a mineral extraction scheme in the Antarctic as a threat to the pristine Antarctic environment and fear that its ecosystem could be destroyed; especially whales, seals and birds. The goals of this group are summarized by Barnes: [95]

"The Antarctic region should be completely protected from all mineral and oil exploration and exploitation. All fishing should take place only as part of a scientific experiment. No actions should be taken by nations, companies or individuals which further jeopardize endangered and threatened species in the region. The region should continue to be demilitarized and no nuclear activities should be allowed. Resumption of commercial sealing should not be allowed, and nor should exploitation of penguins be permitted."

However, the quest for a mineral regime continues. Major guidelines included in recommendation XI–1 state, that the minerals regime should be negotiated by the Consultative Parties in order to maintain the Antarctic Treaty in its entirety and to protect the existing ecosystem. Article IV should be safeguarded and not be affected by the regime. This regime then should cover the entire continent of Antarctica, including the adjacent offshore areas, excluding the deep-sea-bed, while covering every mineral resource at any relevant stage. However, the regime should be open in the sense that it should contain provisions for adherence by states other than the Consultative Parties and cooperation with other relevant international organizations while protecting the responsibilities of the Treaty Parties with respect to the environment of the whole Antarctic Treaty area. The regime should promote necessary research for environmental and resource management decisions that are required.

Reliable and affordable supplies of energy are essential for the economic growth of developing countries. Fifty-nine out of 80 developing countries are net importers of energy. Estimated annual energy investment as a percentage of annual total public investment during the early 1980s reached in many LDC's 40% and above. However, this varied drastically with income level of the individual countries, where very poor countries spent less than 10%.[96] Fifteen percent of foreign assistance was spent on the energy supply sector in LDC's.[97] The World Bank estimated that investments of $125 billion annually (twice the current level) would be needed in developing countries to provide adequate supplies of electricity.[98] According to a World Bank estimate electricity accounts for about 50% of total annual average expenditures on commercial energy supply facilities for developing countries, oil counts for 40% and natural gas and coal 5% each. Investment in a $7.6 billion Antarctic oil field then would be the equivalent of 6% of the above-mentioned $125 billion; this individual oil investment then amounts to more than either the natural gas or coal expenditures.

Financing energy projects depends heavily on the borrowing capacity of LDC's. The poorest countries are highly dependent on concessional aid (which accounted for 80% of their total external borrowing for the energy sector in 1975–1980). Their success in acquiring funds will depend on the extent of the increase in concessional flows. On the other hand, the middle-income countries depend mainly (80%) on export-related and private financial flows for their external financing of energy investments. The situation is particularly acute in highly-indebted developing countries.[99]

Oil consumption increased by 4.5% from 1971 through 1987 in LDC's.[100] Oil consumption is expected to continue rising by about 3%/yr, thereby doubling between 1985–2010. The entire developing world uses about 25% less oil than the United States alone.[99] More than half of the low and lower middle-income countries import 90% or more of their commercial energy; almost all of the imports are in the form of oil.[99] This indicates the import-dependence of resource-lacking LDC's – in particular the poorest LDC's. By applying the common heritage principle to the Antarctic, a modus for resource allocation would have to be established that could guarantee a share for LDC's.

The other question related to petroleum development are the fossil fuel emissions possibly contributing to the Greenhouse effect. Since Antarctic petroleum development is not perceived to happen in the near future the possibility exists that alternative non-fossil fuels will be technically sufficiently developed to make petroleum development in the Antarctic redundant.

However, reducing U.S. CO^2 emissions over the next 50 years much below present levels would require both very large efficiency improvements and the aggressive deployment of non-fossil sources. One without the other will not be efficient.[101] None of the non-fossil energy sources are ready to be substituted competitively for fossil fuels at the scale necessary to reduce CO^2 emissions. A three-pronged R&D strategy is required: improvement of the efficiency of energy conversion and use, improvement of non-fossil energy sources, and improvement of technologies tailored to meet the needs of developing nations. None of the non-fossil energy sources, separately or collectively, are yet ready to be put into use at the level of performance, cost, and social acceptance required to be competitive.[102] Fulkerson et al.[102] estimated that an additional combined public and private sector R&D investment of about $1 billion per year is required. This $1 billion is divided as follows:[102] $300 million per year for to improve the efficiency and economics of end-use and conversion technologies; to improve nuclear power, the additional cost might be $3 billion to $4 billion over the next 10 years or about $350 million per year; to improve solar and other renewables, the additional cost is about $200 million per year; the budgets for biomass, hydroelectric, photovoltaics, solar thermal electric, and wind should be increased by a factor of 2 over several years. Technical and economic potential for fusion could be established in 15 to 20 years, if worldwide efforts of $1 billion to $2 billion annually would be coordinated. An additional $100 million to $20 million annually is required to develop new technologies or to adapt existing ones to the needs of developing nations. A tax on fossil fuel use could raise the public sector portion. A tax rate of as little as 0.2% would raise about $600 million per year.

CONCLUSIONS

No petroleum development will be undertaken in Antarctica in the near future. This is due the following factors: No petroleum reservoirs have been found till present; no production platform exists currently that could withstand the severe environmental forces on a permanent basis and no legal system is in place that could grant mineral rights. The economics certainly are not favorable for any kind of development as demonstrated by the base case and the sensitivity analyses. Protecting the unique Antarctic environment from ecological damage is one of the major concerns, and the liability of a possible oil spill will severely affect the feasibility of such a project. However, the strategic importance of oil could shift the economics in favor of such development, provided oil would be found. This was demonstrated by the Gulf war in 1991, where operation Desert Shield was projected to cost over $8 billion for 1990 alone[103] – more than the required capital investment for this Antarctic development scenario. Non-fossil energy sources still cannot replace the petroleum share in the international energy market. Resource allocation of Antarctic oil would in some way have to consider the needs of developing countries. Due to these unsolved problems Antarctic oil will most likely not contribute to the world's energy supply in the foreseeable future.

REFERENCES

1   U.S. Congress, Office of Technology Assessment, 1989, *Polar Prospects: A Minerals Treaty for Antarctica*, OTA–O–428 (Washington, D.C.: U.S. Government Printing Office)

2   Elliot, D.H., 1985, *Physical Geography – Geological Evolution*, in Bonner, W.N. and Walton, D.W.H., Key environments – Antarctica, Pergamon Press, pp. 39–61.

3   Dalziel, I.W.D. and Elliot, D.H., 1982, *West Antarctica: Problem Child of Gondwanaland*, Tectonics, vol. 1, no. 1, pp. 3–19.

4   Rowley, P.D., 1983, *Developments in Antarctic Geology during the Past Half Century*, in Boardman, S.J., ed., Revolution in the Earth Sciences – Advances in the Past Half Century, Dubuque, Iowa, Kendall/Hunt Publishing Company.

5   Craddock, C., 1982, *Antarctica and Gondwanaland* (Review Paper), in Craddock, C., ed., Antarctic Geoscience, Madison, University of Wisconsin Press, pp. 3–13.

6   James, P.R. and Tingey, R.J., 1983, *The precambrian geological evolution of the East Antarctic metamorphic shield – a review*, in Oliver, R.L., James, P.R. and Jago, J.B., eds., Antarctic Earth Science, Cambridge University Press, Australian Academy of Science.

7   Elliot, D.H., 1975, *Gondwana Basins of Antarctica*, in Campbell, K.S.W., ed., Gondwana Geology, Proceedings of the Third Gondwana Symposium 1973, Australian National University Press.

8   Behrendt, J.C., *Scientific Studies relevant to the question of Antarctica petroleum resource potential*, in Geology of Antarctica, Tingey, R.J., ed., Oxford University Press, in press.

9   St. John, B., 1986, *Antarctica – Geology and Hydrocarbon Potential*, in Halbouty, M., ed., Future Petroleum Provinces of the world, Memoir 40, AAPG, 55–100.

10 Behrendt, J., 1990, *Recent Geophysical and Geological Research in Antarctica related to the Assessment of Petroleum Resources and Potential Environmental Hazards to their Development*, in Splettstoesser, J.F., Dreschhoff, G.A., eds., Mineral Resources Potential of Antarctica, Antarctic Research Series vol. 51, American Geophysical Union.

11 Masters, C.D., Altmasi, E.D., Ditzman, W.D., Meyer, R.F., Mitchell, B., Root, D.H., 1987, *World resources of crude oil, natural gas, natural bituman, and shale oil*, World Petroleum Congress Proceedings 12th, 5, 3–27.

12 Davey, F.J., Bennett, D.J., Houtz, R.E., 1982, *Sedimentary basins of the Ross Sea, Antarctica*, New Zealand Journal of Geology and Geophysics, v. 25, pp. 245–255.

13 Lock, R.G., 1983, *Continental Margin Petroleum Potential in the Ross Sea Region*, New Zealand Antarctic Record, v. 5, no. 1.

14   Houtz, R.E., Davey, F.J., 1973, *Seismic profiler and sonobuoy measurements in Ross Sea, Antartica*, Journal of Geophysical Research, v. 78, pp. 3448–3468.

15   Behrendt, J.C., 1983, *Geophysical and Geological Studies Relevant to Assessment of the Petroleum Resources of Antarctica*, in Oliver, R.L., James, P.R. and Jago, J.B., eds., Australian Academy of Science and Cambridge University Press.

16   Cooper, A.K., Davey, F.J., Hinz, K., 1990, *Geology and hydrocarbon potential of the Ross Sea, Antarctica*, in St. John, B., ed., Antarctica as an Exploration Frontier, The American Association of Petroleum Geologists, AAPG Studies in Geology #31.

17   Cooper, A.K., Davey, F.J., Hinz, K., in press, *Crustal extension and origin of sedimentary basins beneath the Ross Sea and Ross Ice Shelf, Antarctica*, in Thomson, M.R.A., Crame, J.A., Thomson, J.W., eds., Geological Evolution of Antarctica, Cambridge, Cambridge University Press.

18   Fritsch, J., 1980, *Bericht uber geohysikalische Messungen in Ross Meer/Antarktis wahrend der Monate Januar/Februar 1980*, Hannover, Federal Republic of Germany, Bundesanstalt fur Geowissenschaften und Rohstoffe, 36p.

19   Hayes, D.E., Frakes, L.A., 1975, *General Synthesis Deep Sea Drilling Project Leg 28* in Initial Reports of the Deep Sea Drilling Project, California University, Scripps Institution of Oceanography, La Jolla, v. 28, pp. 919–942.

20   Shipboard Scientific Party, 1975, *Initial Reports of the Deep Sea Drilling Project, Part 1*, Shipboard site reports, California University, Scripps Institution of Oceanography, La Jolla, v. 28, pp. 1–369.

21   McIver, R.D., 1975, *Hydrocarbon gases in canned core samples from Leg 28 sites 271, 272, and 273 Ross Sea*, in Initial Reports of the Deep Sea Drilling Project, California University, Scripps Institution of Oceanography, La Jolla, v. 28, pp. 815–819.

22   Claypool, G.E., Kvenvolden, K.A., 1983, *Methane and other hydrocarbon gases in marine sediment*, Annual Reviews Earth Planetary Science, v. 11, pp. 299–327.

23   Kvenvolden, K.A., Cooper, A.K., 1987, *Natural gas hydrates of the offshore circum-pacific margin – a future energy resource?* in Proceedings of Circum-Pacific Council Energy and Mineral Resources Symposium August 1986, Houston Circum Pacific Council for Energy and Mineral Resources.

24   Wright, N.A., Williams, P.L., 1974, Mineral Resources of Antarctica, U.S. Geological Survey Circular 705, 29p.

25   Thomas, B.M., 1982, *Land plant source rocks for oil and their significance in Australian Basins*, APEA Journal, v. 22, Part 1, pp. 164–178.

26   McIntyre, R.D., Wilson, G.J., 1966, *Preliminary Palynology of Some Antarctic Tertiary erratics*, New Zealand Journal of Botany, v. 4, no. 3, pp. 315–321.

27   Cameron, P.J., 1981, *The Petroleum Potential of Antarctica and its Continental Margin*, APEA Journal, v. 21, pp. 99–111.

28   Collen, J.D., Barrett, P.J., 1990, *Petroleum Geology from the Ciros-1 Drill Hole, McMurdo Sound: Implications for the Potential of the Victoria Land Basin, Antarctica*, in St. John, B., ed., Antarctica as an Exploration Fron-

tier, The American Association of Petroleum Geologists, AAPG Studies in Geology #31.

29   Cook, R.A., Davey, F.J., 1984, *The hydrocarbon exploration of the basins of the Ross Sea, Antarctica, from modelling of the geophysical data*, Journal of Petroleum Geology, v. 7, 213–226.

30   Hinz, K., Block, M., 1983, *Results of geophysical investigations in the Weddell Sea nd in the RossSea, Antarctica*, Proceedings of 11th World Petroleum Congress (London), PD2, no.1, pp. 1–13.

31   Cook, R.A., Woolhouse, A.D., 1989, *Residual hydrocarbons*, in Barrett, P.J., ed., Antarctic Cenozoic history from the CIROS–1 drillhole, McMurdo Sound, New Zealand Department of Scientific and Industrial Research Bulletin 245, Wellington, pp. 211–218.

32   Lowery, J.H., 1989, *Coal particles and laminae*, in Barrett, P.J., ed., Antarctic Cenozoic history from the CIROS–1 drillhole, McMurdo Sound, New Zealand Department of Scientific and Industrial Research Bulletin 245, Wellington, pp. 219–222.

33   Rapp, J.B., Kvenvolden, K.A., Golan-Bac, M., 1987, *Hydrocarbon geochemistry of sediments offshore from Antarctica*, in Cooper, A.K., Davey, F.J., eds., The Antarctic continental margin: geology and geophysics of the western Ross Sea, Circum Pacific Council Energy and Mineral Resources, Earth Science Series, v. 5B, Houston, pp. 217–224.

34   Sackett, W.M., Poag, C.W., Eadie, B.J., 1974, *Kerogen recyling inthe Ross Sea*, Antarctica, Science, v. 185, pp. 1045–1047.

35   Cook, R.A., 1988, *Interpretation of the geochemistry of oils of Taranaki and West Coast Region, western New Zealand*, Energy Exploration and Exploitation Journal, v. 6, pp. 201–202.

36   Johnston, J.H., Collier, R.J., Craig, J.T., 1988, *Oil source rock correlations in the Maui-4 oil exploration well, south Taranaki Basin*, Energy Exploration and Exploitation Journal, v. 6, pp. 233–247.

37   Deacon, G.E.R., 1977, *Antarctic Water Masses and circulation*, in Dunbar, M.J., ed., Polar Oceans, Artic Institute of North America, pp. 11–16.

38   Tchernia, P., 1980, *Descriptive Regional Oceanography*, Pergamon.

39   Keys, J.R., 1984, *Antarctic Marine Environments and Offshore Oil*, Commission for the environment, Wellington New Zealand.

40   Steten, N.A., Troup, A.J., 1973, *A synoptic climatology of satellite observed cloud vortices over the Southern Hemisphere*, Quarterly J. Royal Meteorological Society, v. 99, pp. 56–72.

41   Schweidtfeger, W., 1970, *The climate of the Antarctic*, in Landsberg, H.F., ed., World Survey of Climatology, Elsevier, v. 14, ch. 4, pp. 253–355.

42   Parish, T.R., 1980, *Surface winds in East Antarctica*, PhD Thesis, Research Report, Department of Meteorology, University of Wisconsin, Madison, 121 pp.

43   Gordon, A.L., 1980, *Comments on Southern Ocean near-surface circulation and its varaibility*, Annals of Glaciology, v. 1, pp. 57–60.

44   U.S. Navy, 1957, *Operation Deep Freeze II Oceanographic Survey Results*, U.S. Naval Hydrographic Office, Washington, D.C.

45   U.S. Navy, 1963, *Antarctic ice observations October 1962–March 1963*, U.S. Naval Hydrographic Office, Washington, D.C.

46   Ainley, D.G., Jacobs, S.S., 1981, *Sea bird affinities for ocean and ice boundaries in the Antarctic*, Deep Sea Research, v. 28, pp. 1173–85.

47   Jacobs, S.S., Amos, A.F., Bruchhausen, P.M., 1970, *Ross Sea oceanography and Antarctic Bottom Water formation*, Deep Sea Research, v. 17, pp. 935–62.

48   Jacobs, S.S., Gordon, A.L., Ardai, J.L., 1979, *Circulation and melting beneath the Ross Ice Shelf*, Science, v. 203, pp. 439–43.

49   U.S. Navy, 1981, *Marine Climate Atlas of the World*, v. 9, Department of the Navy, Oceanographic Office, Washington, D.C.

50   Cammaert, A.B., Muggeridge, D.B., 1988, *Ice Interaction with Offshore Structures*, Van Nostrand Reinhold.

51   Giannesini, J.F., Champlon, D., Badour, D., 1991, Innovative Technologies Reduce Cost of Offshore Marginal Field Developments in Northwestern Europe, SPE 22028, 35th SPE Hydrocarbon Economics and Evaluation Symposium, Dallas.

52   Iyer, S.H., 1989, *A State of the Art Review of Local Ice Loads for the Design of Offshore Structures*, in Timco, G.W., ed., Working Group on ice forces, 4th state of the art report, U.S. Army Cold Regions Research and Engineering Laboratory, Special Report 89–5.

53   Croasdale, K.R., 1984, *Sea Ice Mechanics: A General Overview*, MTS Journal, vol. 18, no. 1.

54   Bea, R.G., 1984, *Foundations for Arctic Offshore Structures*, MTS Journal, vol. 18, no. 1.

55   Goodman, M.A., 1978, *Handbook of Arctic Well Completions*, Gulf Publishing Co.

56   Watt, B., 1984, *Geotechnical Issues Affecting Offshore Development: An Overview*, MTS Journal, vol. 18, no.1.

57   Bea, R.G., 1984, *Foundations for Arctic Offshore Structures*, MTS Journal, vol. 18, no. 1.

58   Amoco Production Company, 1985, *Exploration Plan for Navarin Basin OCS Sale 83 Area Bering Sea, Alaska*.

59   Gaida, K.P., Barnes, J.R., Wright, B.D., 1983, *Kulluk – An Arctic Exploratory Drilling Unit*, OTC 4481, v. 1, pp. 337–343.

60   Karsan, D.I., Borreson, R., 1991, *Concrete TLP for Sub-Arctic Conditions*, Offshore Technology Conference, OTC 6569.

61   Wilson, R., 1988, *A Review of the Development of the SWOPS Subsea Production System*, OTC 5724, Offshore Technology Conference.

62   Behrendt, J.C., in press, *Scientific studies relevant to the question of Antarctica petroleum resource potential*, in Tingey, R.J., ed., Geology of Antarctica, Oxford University Press.

63   Croasdale, K.R., 1984, *Sea Ice Mechanics: A General Overview*, MTS Journal, vol. 18, no. 1, pp. 8–16.

64   Wessels, E., Kato, K., 1989, *Ice forces on fixed and floating conical structures*, in Timco, G.W., ed., Working Group on ice forces, 4th state of the art

report, U.S. Army Cold Regions Research and Engineering Laboratory, Special report 89–5.

65 Frederking, R., Schwarz, J., 1982, *Model test of ice forces on fixed and oscillating cones*, Cold Regions Science and Technology, vol. 6, pp. 61–72.

66 Matsuishi, M., Ettema, R., 1986, *Model Study of a floating, moored platform in a moving field of mushy ice rubble*, IAHR Ice Symposium, Proc. vol. 1, pp. 197–209, Iowa City.

67 Pilkington, G.R., Wright, B.D., Dixit, B.C., Woolner, K.S., O'Dell, B.D., 1986, *A review of Kulluk's performance after three years operations in the Beaufort Sea*, IAHR Ice Symposium, Proc. vol. III, pp. 145–176, Iowa City.

68 Demirbilek, Z., 1989, *Tension Leg Platform: An overview of the Concept, Analysis and Design*, in Demirbilek, Z., ed., Tension Leg Platform – A state of the art review, American Society of Civil Engineers.

69 Fjeld, S., Rosendahl, F., 1990, *The Application of Concrete for Floating Offshore Structures*, 9th International Conference on Offshore Mechanics and Arctic Engineering, v.3 part A, pp. 105–114.

70 de Oliveira, J., 1988, *Concrete Hulls for Tension Leg Platforms*, Offshore Technology Conference, OTC 5636.

71 Lanan, G.A., Niedoroda, A.W., Palmer, A.C., 1984, *Arctic Hydrocarbon Transportation Systems in the Twenty-First Century*, MTS Journal, vol. 18 no. 1, pp. 45–53.

72 U.S. Dept. of the Interior, Minerals Management Service, 1987, *Oil Spill Response Measures for Offshore Oil and Gas Operations*, OCS Report MMS 87–0062.

73 U.S. Dept. of the Interior, Minerals Management Service, 1985, *Norton Basin Sale 100 Final EIS*.

74 U.S. Dept. of the Interior, Minerals Management Service, 1990, *Beaufort Sea Planning Area Oil and Gas Lease Sale 124 – Final Environmental Impact Statement*, OCS EIS/EA MMS 90–0063.

75 Kirstein, B.E., Redding, R.T., 1988, Ocean-Ice Oil-Weathering Computer User's Manual, Final Reports of Principal Investigators, vol. 59, RU 664, Anchorage, USDOC, NOAA, OCSEAP, USDOI, MMS, Alaska OCS Region, pp. 1–145.

76 Baker, J.M. Clark, Kingston, P.F., Jenkins, R.H., 1990, *Natural Recovery of Cold Water Marine Environments After an Oil Spill*, presented at the 13th Annual Arctic and Marine Oilspill Program Technical Seminar.

77 Payne, J.R., McNabb, G.D., Jr., Hachmeister, L.E., Kirkstein, B.E., and Clayton, J.R., 1987, Development of a predictive model for the weathering of oil in the presence of sea ice, Final report prepared for NOAA, OCS/EIA program, Washington, D.C., U.S. Department of Commerce.

78 Wilson, D.G., MacKay, D., 1986, The behavior of oil in freezing situations, in Prodeedings of the 9th Arctic Marine Oil Spill Program, Technical Seminar, Ottawa.

79 McWhinnie, M.A., 1977, Marine Biology, in Elliot, D.H., ed., A Fremework for Assessing Envirnmental Impacts of Possible Antartic Mineral Development, Ohio State University, Columbus, Ohio.

80   Zumberge, J.H., 1979, Possible environmental effects of mineral exploration and exploitation in Antarctica, Cambridge.

81   Siniff, D.B., 1977, Preliminary assessment: Antarctic biological environment, marine mammals, in Elliot, D.H., ed., A Framework for Assessing Environmental Impacts of Possible Antartic Mineral Development, Ohio State University, Columbus, Ohio.

82   Kennicutt II, M.C., Sweet, S.T., McDonald, T.J., Denoux, G.J., 1991, *Oil Spills in Polar Climates: The Bahia Paraiso Accident*, OTC 6528.

83   Kennicutt II, M.C., Sweet, S.T., Fraser, W.R., Culver, M., Stockton, W.L., 1991, The Fate of Diesel Fuel spilled by the Bahia Paraiso in Arthur Harbor, Antarctica, Oil Spill Conference.

84   Kennicutt II, M.C., et al., 1990, *Oil spillage in Antarctica*, Env. Science and Technology, v. 24.

85   Kennicutt II, M.C., Sweet, S.T., Fraser, W.R., Stockton, W.L., Culver, M., 1991, Grounding of the Bahia Paraiso at Artur Harbor, Antarctica. 1. Distribution and Fate of oil spill related hydrocarbons, Env. Science and Technology, v. 25.

86   Pontecorvo, G., 1982, *The Economics of the Resources of Antarctica*, in Charney, J.J., ed., The New Nationalism and the use of Common Spaces, International Law, pp. 155–166.

87   Potter, N., 1969, *Natural Resource Potentials of the Antarctic*, New York, American Geographical Society, 97p.

88   Cameron, W.L. and G. Pollio, 1987, *Financing Mine Development in Today's Uncertain Environment*, Minerals and Materials, February/March, Bureau of Mines, United States Department of the Interior.

89   Greenwald, J., 1991, *Crude in full retreat*, Time, January 28

90   Beeby, C.D., 1983, *An Overview of the Problems which should be Addressed in the Preparation of a Regime Governing the Mineral Resources of Antarctica*, in Vicuna, O., Antarctic Resources Policy, Cambridge University Press.

91   Zorn, S.A., 1984, *Antarctic Minerals, a Common Heritage Approach*, Resources Policy, v. 10, n. 1.

92   Auburn, F., 1984, *Antarctic Minerals And The Third World*, Fram Journal of Polar Studies, v. 1, n. 1.

93   Zumberge, J.H., 1982, *Potential Mineral Resource Availability and Possible Environment Problems in Antarctica*, in Charney, J.I., ed., The New Nationalism And the Use of Common Spaces, The American Society of International Law.

94   Mitchell, B., Tinker, J., 1980, *Antarctica and its Resources*, Earthscan.

95   Barnes, J., 1982, *Let's Save Antarctica!*, Greenhouse Publications.

96   Munasinghe, Mohan, 1990, *Electric Power Economics*, Butterworths, London.

97   World Bank, 1989, *Annual Report 1989*, Washington D.C.

98   World Bank, 1990, *Capital Expenditure for Electric Power in the Developing Countires in the 1990's*, World Bank Industry and Energy Development Working Paper, Energy Series Paper No. 21, Washington, D.C.

99 Office of Technology Assessment, 1991, Energy in Developing Countries, U.S. Congress, Washington, D.C. Government Printing Office.

100 International Energy Agency, 1989, *World Energy Statistics and Balances 1971–1987*, OECD, Paris.

101 Oak Ridge National Laboratory, 1989, *Energy Technology R&D: What could make a difference? Part 1: Synthesis Report*, Oak Ridge Tennessee.

102 Fulkerson, W, Reister, D.B., Perry, A.M., Crane, A.T., Kash, D.E., Auerbach, S.I., 1989, *Global Warming: An Energy Technology R&D Challenge*, Science, v. 246, November 17.

103 Church, G.J., 1990, Is uncle Sam being suckered, Time, December 24.

104 Cooper, A.K., Davey, F.J., 1985 *Episodic rifting of Phanerozoic rocks in the Victoria Land Basin, western Ross Sea, Antarctica*, Science v. 229, pp. 1085–1089.

105 Cooper, A.K., Davey, F.J., Behrendt, J.C., 1987, *Seismic stratigraphy and structure of the Victoria Land Basin, Western Ross Sea, Antarctica*, in Cooper, A.K., Davey, F.J., eds., The Antarctic continental margin: geology and geophysics of the western Ross Sea, Circum Pacific Council Energy and Mineral Resources, Earth Science Series, v. 5B, Houston, pp. 27–76.

106 Barrett, P.J., 1981, *History of the Ross Sea region during the deposition of the Beacon Supergroup 400–180 million years ago*, Journal Royal Society of New Zealand, v. 11, pp. 447–458.

107 Collen, J.D., Yang, Xinghua, Collier, R.J., Johnston, J.H., 1989, *Hydrocarbon source rock potential and organic maturation*, in Barrett, P.J., ed., Antarctic Cenozoic history from the CIROS–1 drillhole, McMurdo Sound, New Zealand Department of Scientific and Industrial Research Bulletin 245, Wellington, pp. 23–230.

108 Tissot, B., Durand, B., Espitalie, J., Combaz, A., 1974, *Influence of nature and diagenesis on organic matter in formation of petroleum*, AAPG Bulletin, v. 58, pp. 499–506.

109 Mildenhall, D.C., 1989, *Terrestrial palynology*, in Barrett, P.J., ed., Antarctic Cenozoic history from the CIROS–1 drillhole, McMurdo Sound, New Zealand Department of Scientific and Industrial Research Bulletin 245, Wellington, pp. 119–128.

110 Hill, R.S., 1989, *Fossil leaf*, in Barrett, P.J., ed., Antarctic Cenozoic history from the CIROS–1 drillhole, McMurdo Sound, New Zealand Department of Scientific and Industrial Research Bulletin 245, Wellington, pp. 143–144.

111 Collen, J.D., Frogatt, P.C., 1986, *Depth of burial and hydrocarbon source potential*, in Barrett, P.J., ed., Antarctic Cenozoic history from the MSSTS–1 drillhole, McMurdo Sound, New Zealand Department of Scientific and Industrial Research Bulletin 237, Wellington, pp. 161–164.

112 White, P., 1989, *Downhole logging*, in Barrett, P.J., ed., Antarctic Cenozoic history from the CIROS–1 drillhole, McMurdo Sound, New Zealand Department of Scientific and Industrial Research Bulletin 245, Wellington, pp. 7–14.

113 Bridle, I.M., Robinson, P.H., 1989, *Diagenesis*, in Barrett, P.J., ed., Antarctic Cenozoic history from the CIROS-1 drillhole, McMurdo Sound, New Zealand Department of Scientific and Industrial Research Bulletin 245, Wellington, pp. 201–208.

114 Hinz, K., Wilcox, J.B., Whiticar, M., Kudrass, H.R., Exon, N.F., Feary, D.A., 1986, *The west Tasmanian margin: an underrated petroleum province?* in Glenie, R.C., ed., Second southeastern Australia oil exploration symposium, Melbourne, Petroleum Exploration Society of Australia, pp. 395–410.

115 Anderson, J.B., Clark, H.C., Weaver, F.M., 1977, *Sediments and Sediment Processes on high Latitude Continental Shelves*, OTC 2738, Offshore Technology Conference, Houston, TX.

116 Han-Padron Associates, 1985, *Beaufort Sea Petroleum Technology Assessment*, US Dept. of the Interior, OCS Study MMS 85–0002, Technical Report #112.

117 Annonymous, 1988, *Phillips Norway lets contract for Ekofisk tank barrier*, Oil and Gas Journal, February 22.

118 Behrendt, J.C., 1990, *Multichannel Seismic Reflection Surveys over the Antarctic Continental Margin relevant to Petroleum Resource Studies*, in St. John, B., ed., Antarctica as an Exploration Frontier, The American Association of Petroleum Geologists, AAPG Studies in Geology #31.

119 American Petroleum Institute, 1989, *The American Petroleum Institute Task Force Report on Oil Spills*, Washington, D.C.

120 Garrett, J.N., 1984, *The Economics of Antarctic Oil*, in Alexander, L. and Hanson, L., eds., Antarctic Politics and Marine Resources, Critical Choices for the 1980s.

121 Gordon, A.L., 1977, *Southern Ocean physical oceanography*, in Elliot, D.H., ed., A Framework for Assessing Environmental Impacts of Possible Antartic Mineral Development, Ohio State University, Columbus, Ohio.

122 McWhinnie, M.A., 1977, *Marine biology*, in Elliot, D.H., ed., A Framework for Assessing Environmental Impacts of Possible Antartic Mineral Development, Ohio State University, Columbus, Ohio.

123 Mackay, D., McAuliffe, 1989, *Fate of hydrocarbons discharged at sea*, Oil and Chemical Pollut., 5, 1–20.

124 Cameron, R.E., Honour, , R.C., Morelli, F.A., 1977, *Environmental Impact studies at Antarctic sites*, in Llano, G.A., ed., Adaptations in Antarctic Ecosystems, SCAR, Smithonian Institution, Washington, D.C.

125 Parker, B.C., Howard, R.V., Allnut, F.C.T., 1978, *Summary of environmental monitoring and impact assessment of the DVDP*, in Parker, B.C., ed., Environmental Impact in Antarctica, Virginia Polytechnic Institute and State University, Blacksburg, Va.

126 Broady, P.B., Greenfield, L.G., Wynn-Williams, O., 1983 Antarctic Research Unit, University of Canterbury, N.Z, Expedition Report 21.

127 Sexstone, A.M., Everett, K., Jenkins, T., Atlas, R.M., 1978, *Fate of crude and refined oils in North Slope soils*, Arctic, v. 31, 339–47.

128 Colwell, R.R., Mills, A.L., Walker, J.D., Garcia-Tello, P., Campos-P.V.,

1978, *Microbial ecology studies of the Metula spill in the Straits of Magellan*, J.Fish.Res. Board Can., 35, 573–80.

129 Arhelger, S.D., Robertson, B.R., Button, D.K., 1977, *Arctic hydrocarbon biodegredation*, in Fate and Effects of Petroleum Hydrocarbons in Wolfe, D., ed., Marine Ecosystems and Organisms, Pergamon Press, New York.

130 Haines, J.R., Atlas, R.M., 1982, *In situ microbial degradation of Prudhoe Bay crude oil in Beaufort Sea sediments*, Mar. Env. Res., 7, 91–102.

131 Atlas, R.M., 1985, *Effects of hydrocarbons on micro-organisms and petroleum biodegradation in Arctic exosystems*, in Engelhardt, F.R., ed., Petroleum Effects in the Arctic Environment, London and New York, Elsevier Applied Science Publishers.

132 Cormack, D., 1983, *Response to Oil and Chemical Marine Pollution*, London and New York.

Seasonal Ice Storage for
Cooling Loads
The Icicle Growth Enhancement Method

DWAYNE S. BREGER

Department of Mechanical Engineering
University of Massachusetts
Research Supervisor: Dr J. Edward Sunderland

SUMMARY

This paper describes the design, experimental, and analytical development of a seasonal storage ice system to meet building, agricultural, or process cooling loads. Ice is made during winter using ambient temperature to freeze water in the form of icicles. The icicle growth is enhanced by providing a long vertical surface around which the icicle is formed. Once fully formed, the icicle is released and drops into an excavated pit to form an ice pile. The ice pile is covered after the winter, or the entire system is permanently enclosed and, in the winter, air is brought in through forced ventilation. Melt water from the ice pile is distributed to meet cooling loads throughout the year; return water is sprayed on the ice pile to transfer the building heat to the ice.

The system design is developed in detail along with a summary of other ice storage designs and demonstration projects. The design and construction of an experiment to determine icicle growth rate and effect of design conditions is described. The experimental results are presented but are too limited to fully assess the proposed icicle design performance. Nonetheless, the results are compared with other winter ice growth methods and are encouraging. Cost goals for the icicle design are analyzed in terms of displaced conventional costs and appear to provide feasible economic prospects for the ice system construction and operation.

1. INTRODUCTION

1.1 BACKGROUND

The use of energy for cooling of buildings, agricultural products, and other processes represents a substantial energy use in the United States. Cooling is normally accomplished using electric power as the energy source to drive vapor compression cooling machinery. Therefore, the demand to remove low-temperature heat (low-grade energy) from buildings requires electric power (high-grade energy), resulting is a very poor second law efficiency and questionable management of finite energy resources. Moreover, summer cooling loads often define the electric utility sector peak capacity requirements resulting in operation of peaking facilities and demand for construction of additional electric capacity.

The use of thermal energy for space cooling using absorption or desiccant cycles is an alternative. These technologies are presently available though further development is necessary. The use of solar energy as input for thermal cooling cycles is appealing due to the coincidence of the energy source and load. However, further solar advances are required before solar is expected to be competitive with gas driven cycles.

The development of solar energy for cooling applications has concentrated on Sunbelt applications. Buildings in the northern latitudes in the U.S. also demand significant cooling, particularly for commercial and industrial facilities. This substantial demand also defines the peak capacity of many northern utilities. Although the northern latitudes can also make use of the thermal cooling cycles being developed, another resource is available which offers tremendous savings in energy utilization and environmental impact. The resource is the cold winter temperatures which can be compactly stored (in the form of ice) and used during summer months to provide the cooling of buildings.

Seasonal ice storage is not a new concept, though relatively little attention has been focused on the idea. Several system designs have been developed and technically demonstrated rather successfully. These systems represent a very limited number of possible technical approaches.

1.2 APPLICATIONS AND ENERGY IMPLICATIONS

The difficulty of storing thermal energy over a long period of time can be largely overcome by storing a volume large enough so that the surface area through which heat loss occurs is small relative to the stored volume. A large volume provides economy of scale both in terms of thermal efficiency and the cost of the containment structure. Seasonal ice storage, therefore, must be applied on a relatively large scale. Applications include commercial buildings, supermarkets, industrial or office parks, shopping centers, cooling for the agricultural and food processing industry, or multi-unit residential developments with district cooling networks.

Seasonal storage systems might be owned and operated by the building developer or owner, though more likely a micro-utility may be established using third-party financing to implement such a system. Alternatively, electric utilities may consider investment in such projects as a means to shave peak power demands.

The ice that is used to meet the building cooling load displaces electric energy required to operate a conventional vapor compression cooling system and subsequently the primary fuel burned at the power generating station. A seasonal ice storage system can be designed to meet the entire annual cooling load of a site or may be operated as a base capacity cooling facility, or as a peaking facility. The optimum strategy will be determined by factors such as electric rate structure, load profile, and constraints on the ice storage size (due to space or capital limitations).

The energy purchased during the process of ice formation, storage, and utilization will be small relative to the cooling energy supplied. Present methods for seasonal ice storage achieve an overall coefficient of performance (COP) in the range of 10–30. COP is the ratio of the cooling energy supplied to the total auxiliary electric energy used. Conventional cooling equipment operates with a COP of around 3.

Melt water from the ice storage is used to meet the building cooling load. The water is available at 0°C (32°F), compared to 7°C (45°F) for the chilled water from conventional cooling systems. At the lower temperature, condensation of the building humidity occurs more easily and the allowable comfort zone is more readily maintained even at higher air temperatures. This advantage of the ice system will conserve the total cooling energy demand of the buildings and can reduce the time when conventional cooling systems are operated only for humidity control. Retrofit of building cooling systems to use the ice cooling source requires plumbing the ice system chilled water directly into the building chilled water distribution piping.

On a national basis the potential for energy savings is very large. Although the system applicability is limited to locations with sufficient freezing ambient winter temperatures and land area, the availability of suitable existing and new building stock in these regions is enormous. The applicable geographical regions include New England and New York, the Great Lakes region, and the Northern and Mountain States, as well as Canada.

1.3   PREVIOUS METHODS FOR ICE STORAGE AND LITERATURE REVIEW

1.3.1   Ice Storage Methods

Historical Perspective: As recently as the first half of the present century, seasonal ice storage was a major industry in New England. During winter and after the surface of lakes were frozen, the ice thickness was increased by continually adding thin layers of water to the ice surface. Man and horse power were used to saw the ice surface into large blocks and haul the ice blocks to carriages and trucks on the shore. Winter ice was distributed to large warehouses for seasonal storage in New England, and by ship to the entire Atlantic coast and to the Caribbean and Europe. From the large storage centers, ice blocks were distributed to consumers for commercial and household ice boxes. This thriving industry met its demise with the advent of modern refrigeration technology.

Off-Peak (Diurnal) Ice Storage: The use of ice to meet building cooling loads is presently an available technology from a rather new and rapidly growing industry. The technology involves the daily formation of ice during periods of off-peak electric rates (night time) and the use of the ice for cooling to reduce the energy use and peak demands during periods of peak electric rates (day time). The cost of the product is returned to the investor by reduced electric peak demand and energy charges and, often, additional incentives by the utility for peak kW savings.

The major benefit of this technology to the U.S. is the potential to reduce the demand of peak electric generating capacity. However, the technology does not conserve energy (kWh's) or alleviate environmental damage and, in fact, is a net consumer of electric energy due to process and storage inefficiencies. That is, the building will purchase additional kWh's of electricity, though the total cost will be lower due to reduced (off-peak) energy and demand charges.

The use of off-peak ice production technology has been combined with annual storage.[1] In this project located in North Carolina, a small conventional ice making system is operated continuously during the year and ice is stored in an excavated pit and used for summer cooling of agricultural produce before shipping. The annual storage completely displaces a large peak summer demand.

Existing Seasonal Ice Storage Methods: Four major seasonal ice storage system designs have been developed and demonstrated. The designs involve very different technical approaches with respective advantages and disadvantages as discussed below.

Princeton University/Prudential Insurance Company – Snow Pond:[2,3] This system uses conventional snow-making machines to fill a large excavated pit with snow or small particles of ice. During snow production, the large dome cover is raised about 10 feet off the ground to provide air circulation to freeze and distribute the snow. During the storage months the insulating dome cover is lowered over the snow pile to protect the ice. The system began operation in Princeton, New Jersey in 1982 to provide cooling for a large modern office building. The system was recently shut down mainly due to high manual operation and maintenance costs.

The system allows for a large amount of snow to be produced. More snow can be made over a winter by adding additional snow machines though a height limit is reached due to the restriction of air circulation under the raised dome. The measured density of the snow was 63% of solid ice so a relatively large volume is needed. System control has not been sufficiently automated and an operator is needed to oversee the system during snow-making periods. A substantial amount of energy is needed to operate the snow machines and an overall COP of about 10 is obtained, which is lower than other ice storage methods.

Illinois State University – Saturated Earth Ice System:[4] This technique requires the excavation of a large volume of earth. The pit is then insulated and lined with a water-proof membrane and refilled with water saturated earth and layers of heat exchange pipes. The top is insulated and covered. Anti-freeze is circulated through the storage and heat is extracted to the ambient winter air until the volume is frozen. The experimental system has been successful and cost- effective applications are proposed for commercial buildings. The scheme considers the saturated earth storage to be constructed underneath a parking lot and thus does not require any land area. This scheme could be done for a new construction site, though a retrofit application would be difficult. Sufficient depth of easily excavated earth is required and the amount of ice energy stored per unit volume is about 50% that of solid ice. Frozen earth first occurs around the heat exchange piping which subsequently reduces heat exchange effectiveness. The expected system COP is in the range of 10–30.

Public Works Canada – Ice Block:[5] A couple systems have been demonstrated in Canada based on the freezing of a solid block of ice by successively freezing thin layers of water. Winter ambient air is drawn into the enclosed storage by fans and blown over the ice surface; water is periodically added (2 mm thickness) to evenly cover the surface. The system allows for overall simplicity and automated operation with a COP over 20. The limiting factor is the height of the ice block which can be grown, which is about 7.5 m (25 feet) in the Canadian location of Quebec City. Large quantities of ice can be made but only by increasing the surface area of the ice block. A large amount of land area would thus be required for a large scale application. The flat surface of the block presents a very limited surface area per land area for heat transfer. In addition, a large forced air circulation system is needed to provide a sufficient cold air supply.

Snowmax Ice Pond:[6] An experimental project was operated during the winter of 1989–90 in Corfu, New York to provide process cooling to a cheese factory. The project involved the New York Energy Research and Development Authority and Snowmax Technologies in Rochester. The system produces winter ice by spraying water from nozzles into the ambient air. Introduced into the spray is a biological ice-nucleator agent patented by Snowmax and currently used to enhance the productivity of snow making in ski resorts. The agent increases the ice production and allows ice generation at marginal freezing temperatures. The density of ice in the storage averaged 0.75 of solid ice and the ice pile was covered by an insulating blanket after the winter production. The project operated successfully during the first year and is still under evaluation.

1.3.2   Literature Review of Heat Transfer Analysis of Freezing Processes

The heat transfer literature on the phase change of water to ice is extensive. However, the specific application of icicle growth is not well addressed. Most of the appropriate literature is directed at inward ice formation within a cylinder which is placed in a sub-freezing ambient condition, or outward ice formation on the outer cylinder surface with a sub-freezing coolant circulated within the cylinder. These conditions are addressed analytically.[7,8,9,10] Experimental results are reported.[11,12,13]

These papers offer useful insight into the phase change phenomenon but are dissimilar to the icicle formation in a significant way. Under the conditions assumed by the above references, ice formation occurs between the coolant and freezing liquid, so that the ice becomes a thermal resistance retarding further heat transfer. As the ice thickness increases, the necessary heat conduction through the solid ice decreases the heat transfer effectiveness and eventually limits the final ice thickness which occurs as a steady state. This process is similar to conventional diurnal ice storage technologies and ice machines.

The icicle formation occurs under the condition that the coolant (ambient air) and freezing liquid are always in direct contact. The thin film of water over the already frozen ice poses a negligible thermal resistance due to conduction. Applications of such conditions have been found in the literature in relation to meteorological phenomenon and, in particular, problems encountered with ice accretion on power lines, antennae, aircraft, and ship rigging.

A broad presentation is provided of phase change heat transfer problems in a text book[14] which specifically addresses conditions in cold climates. One section (page 432) addresses the outward ice growth on a cylinder and analytical methods are presented for several simplifying conditions. A symposium proceedings[15] presents a wide range of recent experiences and applications of cold climate heat transfer problems. The papers from this symposium[16,17] address ice accretion on cylindrical bodies in particular and present useful analytical techniques and results.

Although ice accretion models share the greatest similarity with the icicle heat transfer process, the accretion is based on many small droplets of ice and water impinging on the surface under relatively high wind speeds. This may be appropriate for the option of using a water spray to feed the icicles, but is dissimilar to normal icicle formation where a thin film of water flows down the surface.

2. DESCRIPTION OF ICICLE ENHANCEMENT METHOD

2.1 DISCUSSION OF CONCEPT

Winter coldness is stored by freezing water to take advantage of the large thermal storage capacity provided by the phase change (heat of fusion). Water is used due to its availability, low cost, benign environmental impact, appropriate phase change temperature, and large heat of fusion. In order to freeze water, heat must be extracted from the liquid phase at 0°C (32°F). The heat transfer rate strongly depends on the surface area exposed to the cooling source.

Ice formation is accomplished through the enhanced growth of "icicles" suspended in an array above the storage site. Fully-formed icicles will drop and, over the winter, form an ice pile below. The growth of the icicles is enhanced by vertically hung wire, chain, rope or flat surface along which the introduced water will flow and freeze. The icicles will drop when their own weight exceeds the bond force between the ice and enhancement surface or by periodic heating of the enhancement surface to break the bond. The dropped icicles will break on the ice pile to provide a high packing density. A fine spray of water (and excess water dropping from the icicles) will be added to the ice pile surface and frozen to increase the ice density. The design provides a very large surface area over which the ice can form. The icicle freezes from the inside out, so the critical region of heat transfer remains in direct contact with the ambient air.

The amount of water that can be frozen on an icicle is dependent on the meteorological conditions of the location. The total volume of ice produced in a climate can be controlled by the number and length of icicle formations. The density of the icicle array (number and length of icicles per storage or land area) is, however, constrained by the need to maintain sufficient airflow for effective heat transfer.

An illustration of the proposed ice system (not to scale) is shown in Figure 1. The essential components of the ice system design are discussed next.

2.2 ICE STORAGE AND UTILIZATION

The icicle formation region is suspended over an excavated earth pit which is lined with a water-proof membrane. The excavated earth is bermed around the pit to increase its depth. The ground below and on the sides of the pit may be insulated.

Melt water from the ice pile will collect in the pit. During ice production periods, liquid water (if present) will be pumped from the pit and added to the water supply for ice production. During periods of cooling demand, melt water from the pit is pumped through buried insulated pipes to the building(s) and into the chilled water distribution, where it is heat exchanged with the building air in the HVAC system. The ice water is heated by the building air and

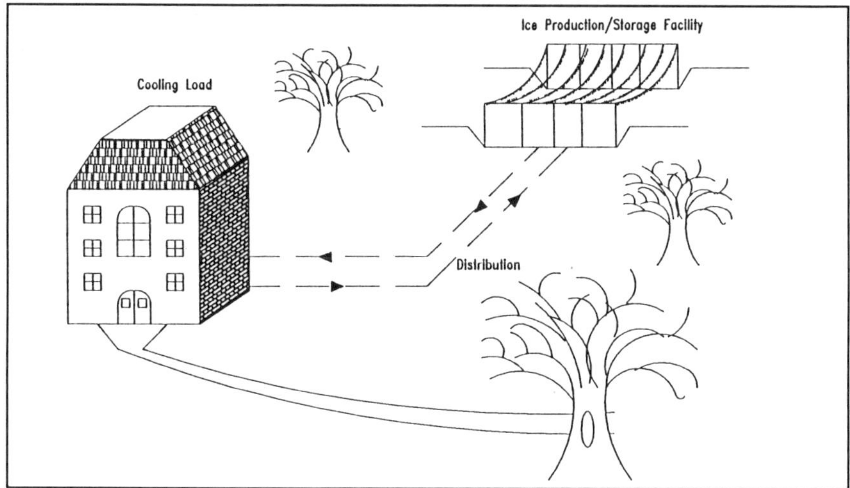

Figure 1. Illustration of Ice System Concept and Application

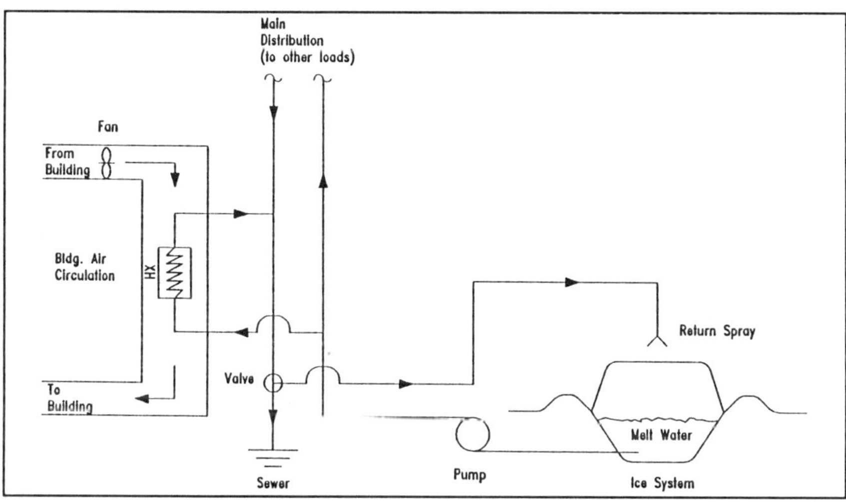

Figure 2. Schematic of Melt Water Distribution System

returned to the ice pile. The return water is sprayed on the ice pile to cause additional melting and transfer the building heat to the ice. A schematic for the system is shown in Figure 2.

During periods of ice production, the ice pile is exposed to the ambient freezing environment. During warmer weather, the ice pile must be protected and insulated from the environment. The ice pit and suspended icicle region can be covered with a movable insulated domed cover similar to that employed in the Princeton/Prudential snow pile system. Alternatively, the entire system could be constructed in a permanent enclosure and ambient air forced through the ice formation region.

2.3 ICE PRODUCTION

Ice production is accomplished by enhancing and controlling a process ordinarily observed in nature that of icicle growth by liquid water flowing down a vertical ice surface leaving a thin water film which freezes, thereby increasing the length and thickness of the ice formation. Icicle formation is enhanced by hanging a series of vertical surfaces such as wire, chain, rope, or thin flat surfaces from parallel cables secured to a structural support along two opposite ends of the pit. For a given climate, annual ice production is controlled by the number and length of these icicle wires. An illustration of this design is shown in Figure 3. The behavior and effectiveness of this process are the main subjects of the experimental work discussed in Section 3.

As water is introduced to the ice production system, the liquid will tend to adhere and flow down the solid surfaces or existing icicles and freeze, forming thicker and longer icicles. Many icicle enhancement surfaces are possible and a sample was included in the experimentation. Multiple wire or flat surface designs are a possibility to increase the surface area and enhance the initial growth rate. Total surface area increases as the icicle size increases.

The full weight of each icicle is supported by the adhesion of the ice around the suspended surface area. The selection of the enhancement surface characteristics is very important so that this adhesive strength is balanced by the desired fully formed icicle weight for automatic release. Alternatively, a heat wire can be attached or enclosed within the enhancement surface and automatically controlled to release the icicle.

A controlled quantity of water is introduced to the ice production process in either a continuous or periodic flow. Water is obtained from any melt water available in the pit or from the main distribution. The water temperature should be just above 0°C (32°F) before being supplied to the ice production. Cooling the supply water to 0°C may be accomplished with a heat exchanger with the ambient air to remove the sensible heat.

Several alternative water supply methods have been considered. Water droplets can be sprayed randomly over the entire icicle production area. Depending on the density of the icicle array, a fraction of the water will settle on the icicles and freeze. The remaining fraction will land on the ice pile and freeze or trickle into the pit. This design provides simplicity and relatively few water outlets. In order to be effective, the water droplets must be properly sized and deployed with appropriate velocity so that sufficient water reaches and adheres to the surfaces and covers most of the available icicle area. A different design for the water supply provides much greater control over the water placement. Flexible piping can be extended along the length of the cables from which the icicle wires are suspended and small holes or nozzle provided to coincide with the icicle wire placements. This approach was used in the experimentation discussed in Section 3.

An alternative to suspending the icicle production over the storage pit is to produce the ice remotely and collect and transport the ice to a central storage facility. The ice collection and transport would have to be highly mechanized. The remote production could be spread out on unemployed winter land area nearby and avoid the structural requirements to suspend the ice production high above the storage pile.

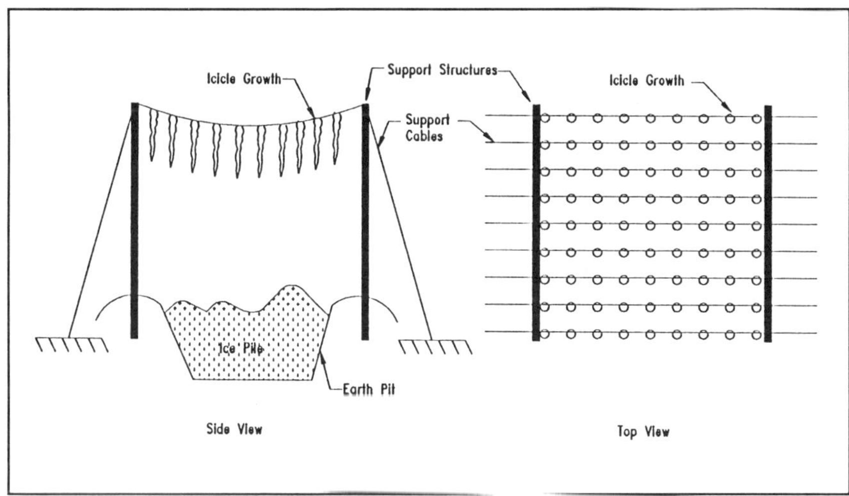

Figure 3. Schematic of Ice Production Technique

2.4 THERMODYNAMIC CONSIDERATIONS

Two major processes define the thermodynamic effectiveness and limitations of the process: (1) the local process surrounding each icicle, and (2) the global process around the entire ice production region.

The local ice formation process involves the introduction of water at a temperature just above freezing. The surrounding air temperature is below the water temperature so heat will be transferred to the air from the water as it flows down the ice surface. Sensible heat will first be transferred until a 0°C (32°F) water temperature is reached; followed by the heat of fusion as the water changes phase into solid ice. The ice bonds with the previous layer of ice or the enhancement surface. Ideally, water is introduced at the top of the icicle such that the thin film of liquid covers the entire icicle surface and just reaches the entire length of the icicle, so that the heat transfer surface area is maximized.

As the water releases the heat of fusion, the ambient air is heated. This heat transfer occurs continuously around each icicle to produce a global heat transfer process which places limitations on the effectiveness and maximum productivity of the ice production scheme. Obviously, the ambient air needs to be replenished at a sufficient rate for useful ice production. Ambient air is either naturally circulated by the ambient wind or can be controlled through forced ventilation. The heated air in the ice production region will tend to rise and a passive air circulation may be induced. As the density of the icicle production increases, air flow may become restricted and limit the heat transfer.

3. ICICLE GROWTH EXPERIMENTATION

3.1 PURPOSE

The experimentation was performed to closely observe the process and problems associated with controlled icicle growth and determine the rate of growth which could be obtained under varying conditions. Conditions studied included the ice enhancement surface, water supply flow rate and temperature, and ambient temperature and wind.

The purpose of the observation of the icicle growth was to improve the system design to increase the icicle growth rate and operation of the system. The results of ice growth rates were to be used in assessing the full scale design specifications and economics of the icicle growth enhancement method.

3.2 DESIGN OF EXPERIMENTAL APPARATUS

The apparatus was designed and constructed to perform the experimental work under outdoor ambient conditions during the winter of 1990–91. The experiment was designed to test the icicle growth behavior and is not concerned with the annual storage aspects which are well established from previous projects (for example,[1,2,6]). The apparatus allows for simultaneous testing of up to 9 icicles so that varying design parameters could be tested under equivalent ambient conditions. A schematic of the experimental apparatus is shown in Figure 4 and a photograph of the constructed project is shown in Figure 5.

3.2.1 Structural Support Tower

The icicles are suspended 14 feet above the ground from a 10 foot by 10 foot wood frame support tower resting on four concrete footings at each corner. The structure includes cross bracing for structural support and scaffolding members to allow safe access to the ice formation and water supply region for experimental control and data collection. The 9 icicles are suspended from 3 2x6 inch cross beams in a 3x3 grid with approximately 2¾ foot spacing.

3.2.2 Water Supply

Water was delivered to the experiment site from a domestic water supply through ½ inch flexible polyvinyl tubing. Hot water could be mixed with the cold main water to regulate supply temperature or to thaw frozen sections. At the experimental site, the main water flow could be diverted to a 25 foot ¼ inch copper coil immersed in an ice bath to reduce the supply temperature. Salt was added to the ice bath to reduce its freezing temperature.

The main water supply flows through two stages of manifolds to split the flow into 9 separate ¼ inch polyvinyl tubes which travel up the support tower and along the icicle support beams to each icicle location. The manifolding

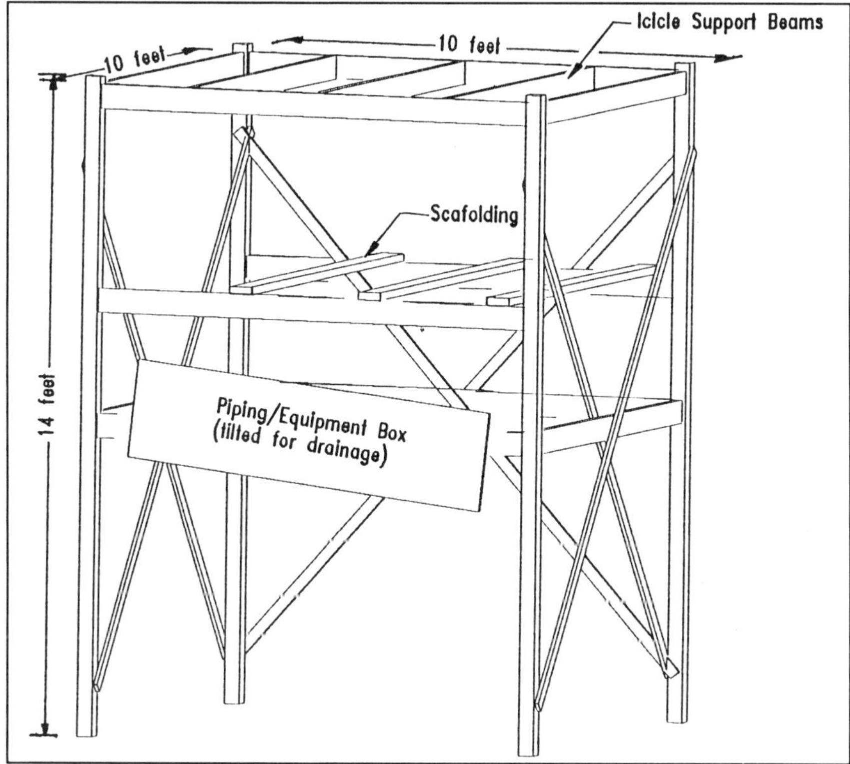

Figure 4. Schematic of Experimental Icicle Support Structure

system and other instrumentation is enclosed in a constructed weatherproof box attached to the side of the tower. The tubing layout was carefully installed so that the main supply and individual icicle tubes would all drain back upon opening the drain valve for freeze protection during non-operating periods.

Water flow rate is controlled at several stages. A ball valve controls the main supply flow to the experiment. Screw clamps after the first manifold can be used to regulate the flow rate to each row of icicles or shut down an entire row. After the second manifold, the flow rate to each of the 9 icicle supply tubes are individually controlled with screw clamps.

Water flow from the ¼ inch icicle supply tube onto the icicle surface was carefully designed to minimize splashing and potential for freeze ups and to allow for manual measurement of the flow rate during experimentation. A photograph of one icicle location is shown in Figure 6. The supply tube is inserted into a small segment of larger tubing which holds it in place and controls the flow down the ice surface. Pulling the supply tube out of the tube segment allows for flow measurement.

Figure 5. Photograph of Experimental Apparatus

Figure 6. Water Distribution to Icicle Enhancement Surface

3.2.3 Icicle Enhancement Surfaces

A variety of icicle enhancement surfaces were selected from local hardware stores. The nine surfaces include various types of large and small gauge chain and a thick and thin diameter polypropylene rope. Each surface was 12 feet (3.6 m) long and hung from the 2x6 inch wood cross beam with a large screw hook and metal ring. The metal ring held the water supply tubing and could be easily lifted off the screw hook to weigh the icicle during experimentation. One location was fitted with a multi-chain surface. The surfaces tested are shown in Figure 7.

As seen in Figure 6, a transition region was designed at the top of the icicle to improve the flow pattern of the water supply to the outside of the icicle. The transition zone is made from aluminum foil wrapped around the surface. Silicone caulk was applied at the top of the region to seal the aluminum foil with the chain or rope to keep the water from adhering to the surface and running through the middle of the aluminum foil.

3.3 Experimental Design

Ice growth experiments were to be run when ambient temperature was expected to stay below about –4°C (25°F) for at least 10 hours. Flow rates were adjusted so that only a small amount of water could be observed dripping off the bottom of enhancement surface. The supply water temperature was noted from the thermocouple placed before the first manifold and flow directed to the cooling coil as necessary.

During the course of the experiment, close visual observations were made of the ice growth behavior and differences between the various surfaces. Data collection was to be made at periodic intervals of about 1 hour. Data collected included ambient temperature from a mercury bulb thermometer, water supply temperature from the thermocouple, and qualitative observation of wind conditions. Flow rates were to be recorded directly for each icicle location by measuring the time necessary to fill a known volume from the icicle supply tube. The weight of each icicle was to be recorded standing on a strain gauge bathroom scale with digital display placed on the tower scaffolding and lifting the full weight of the icicle. (Tare weights of each chain and rope were recorded initially.) Raw data were recorded and analyzed on a computer spreadsheet.

3.4 Experimental Results and Analysis

The results available from experimentation are, unfortunately, very limited and inadequate to fully assess the effectiveness of the ice production method. Several important visual observations were made and data were recorded for one period of operation before the end of winter. System operation during the

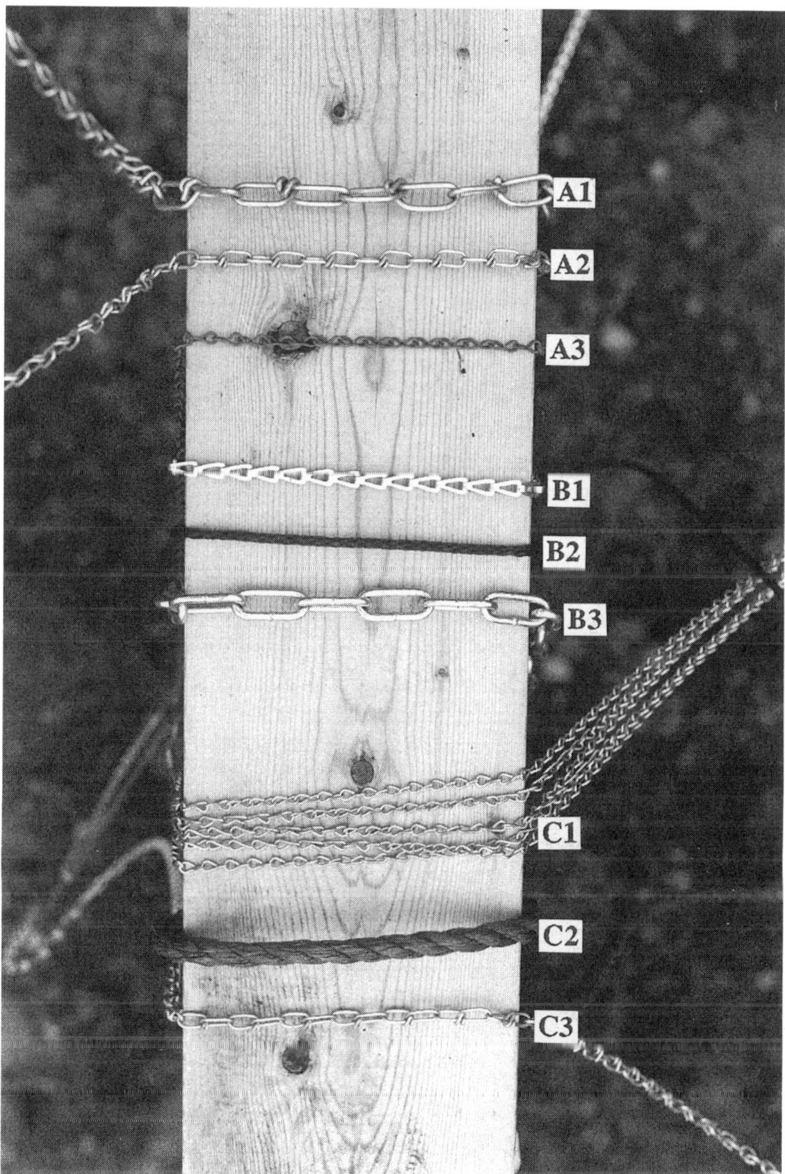

Figure 7. Icicle Enhancement Surfaces used in Experiment

experiment was believed to be greatly sub-optimal; however, the results are encouraging and are compared with another ice storage project[5] in Section 4.

3.4.1 Discussion of Observations

A major problem encountered was the lack of careful control of the supply water temperature. Despite anticipated problems of freezing in the supply tubes, the supply water (from the town main) was reaching the experiment at 4.4°C (40°F), though a temperature of 0–2°C (32–35°F) is necessary for the icicle supply. The copper coil submerged in the salt water ice bath was added to the flow circuit and initially reduced the water temperature to 2°C (35°F); but within an hour the ice bath was warmed and the icicle supply temperature raised to 4°C (39°F). A sufficient supply of ice to add to the bath was not available at the time.

Initial growth of the icicle was observed to be successful. However, after the initial formation of an icicle of about 3.8 cm (1½ inch) diameter, a problem was regularly encountered, especially with the chains. The supply water would adhere to the chain and channel through the entire length of the icicle. In this manner, the supply water was insulated from the ambient air and no further freezing occurred. The transition zone installed at the top of the chains eliminated the problem while the supply temperature was still near freezing. However, at the higher supply temperature, the ice growth did not occur until well below the transition zone and the channelling was frequently observed. The problem of controlling water supply temperature to 0–2°C is not considered serious for a final full scale design, but did present an unexpected problem which was very time consuming to resolve on this experimental level.

Another significant observation was of the flow of the water film along the surface of the icicle. The water film tended to follow a specific trail down the ice surface so that only a small fraction of the available surface area was being wetted. The trail would shift over time so that a roughly cylindrical icicle was formed, but clearly the total surface area available for heat transfer was not effectively being used. A careful design of a supply nozzle may reduce this problem but once on the icicle the water may still amass. The design option of supplying the water through a spray or mist might substantially avoid this problem and lead to a increased rate of ice growth.

The enhancement surface on which the icicle forms did not seem to cause a significant difference in total ice growth. During successful operation, the enhancement surface was covered remarkably quickly and subsequently all icicles had similar surface area characteristics. The use of more complex enhancement surfaces, such as multiple chains or thin flat surfaces, may hold some advantages. The problem of water flow channelling through the center of the icicle was most prevalent with the chains as opposed to the ropes. This is likely due to the openness of the chain links through which the channel can form; though under proper control this may be avoided.

3.4.2 Experimental Results

Only one experiment was run before the end of winter. The experiment was run over a night during the beginning of March. The experiment lasted 8 hours with data taken at the beginning, after 1 hour, and after 8 hours. All nine icicle locations were in operation with different water supply flow rates. The ambient temperature was −3.9°C (25°F) at the beginning of the experiment and dropped to −7.2°C (19°F) during the night until the experiment was stopped. During this time, the temperature of the supply water ranged from 1.7°C (35°F) to 4.3°C (40°F). The ambient wind was generally calm during the experimental period.

For each icicle, the supply flow rate and ice weight were measured at the start and after 1 and 8 hours; therefore, two trial periods of 1 and 7 hour duration were recorded. The results are shown in Table 1 for the 9 icicles. The results show the average ambient conditions over each trial and the experimental results of flow rate and net weight of the ice growth. As discussed in Section 4, the ice growth rate is calculated as the volume of ice grown per degree hour of freezing (DHF) per meter of icicle length (m^3/DHF/m). The effective icicle length (after the transition zone) is 3.35 m (11 feet). This unit of ice growth rate allows for the transformation of results to other climatic conditions and to other icicle lengths and design densities (icicles per area of storage). The growth rate is compared in Section 4 with the ice growth obtained in the Canadian ice block project.[5]

Table 1. Experimental Results of Icicle Growth

		ICICLE LOCATION								
		A1	A2	A3	B1	B2	B3	C1	C2	C3
TRIAL #1										
DURATION	1.00 hr									
T_{amb}	-4.17 °C									
T_{supply}	2.97 °C									
DHF	4.17									
FLOW RATE	liter/min	0.129	0.142	0.179	0.081	0.056	0.169	0.414	0.088	0.264
ICE GROWTH kg, or $10^{-3}m^3$		1.361	0.454	0.000	0.454	0.454	0.000	0.454	0.454	0.454
m^3/DHF/m (x10^{-5})		9.741	3.247	0.000	3.247	3.247	0.000	3.247	3.247	3.247
TRIAL #2										
DURATION	6.92 hr									
T_{amb}	-5.83 °C									
T_{supply}	3.81 °C									
DHF	40.35									
FLOW RATE	liter/min	0.113	0.135	0.142	0.065	0.040	0.153	0.276	0.064	0.169
ICE GROWTH kg, or $10^{-3}m^3$		2.268	0.000	-0.454	2.268	2.268	0.000	0.454	4.082	-0.907
m^3/DHF/m (x10^{-5})		1.677	0.000	-0.335	1.677	1.677	0.000	0.335	0.010	-0.071

The results show a wide range of ice growth and mixed success of the experiment. The maximum ice growth of 9.74 x 10^{-5} m^3/DHF/m occurred during the first hour on icicle A1; this rate is substantially higher than any others recorded and may in part be due to measurement error. However, the results indicate consistent ability of obtaining an ice growth rate of 3.25 x 10^{-5} m^3/DHF/m which occurred on several icicles during the first hour and was nearly sustained on icicle C2 for the full 8 hour duration. The recorded growth rates are considered substantially below the achievable rate due to the observations noted in Section 3.4.1. The water supply was warmer than desired, water channelling through the center of the form icicle was regularly observed, and water supply did not satisfactorily cover the available icicle surface area.

4. ANALYTICAL EVALUATION OF ICICLE ENHANCEMENT SYSTEM

4.1 ESTIMATION OF ICICLE PRODUCTIVITY

The annual capacity for ice production in a given climate is determined by the Degree Hours of Freezing (DHF) which is defined as the cumulative product of the degrees centigrade below 0°C and the number of hours of duration of the temperature. The Canadian experience (discussed in Section 1.3.1) with the Ice Block project[5] can be extended to evaluate the achievable enhanced icicle growth rate which can be compared to the experimental results. Operation of the Canadian Ice Block system indicates that a layer of 2 mm of water can be frozen in 400 DMF (degree minutes of freezing) or 6.67 DHF. Therefore, the ice production rate is (.002 m^3/6.67 DHF) 0.0003 m^3/DHF per m^2 of heat transfer surface area.

The surface area for heat transfer on an icicle increases over time while the icicle grows. Observation of the icicle growth during the experiment showed that the initial growth of ice covering the cold enhancement surface occurred very rapidly, especially for the large gauge chain. If we assume the icicle grows from an initial radius of 0.0125 m (or that the enhancement surface has this radius) to a final radius of 0.05 m, this implies a variation in surface area from 0.079 m^2 to 0.314 m^2 per m of icicle length. If we simply, but conservatively, assume that the radial growth of the icicle is constant over time, then the average surface area over time is the average of the initial and final values; or 0.196 m^2 per m icicle length. Applying the recorded Canadian ice growth rate of 0.0003 m^3/DHF per m^2 surface area to the average size icicle, the expected ice growth rate would be 5.89 x 10^{-5} m^3/DHF per m of icicle length. This analysis assumes that the entire icicle surface area can be wetted by a thin film; although this was not observed in the experimentation, with an improved water supply design (for example, a spray system) this performance is believed to be feasible.

The maximum ice growth rate recorded during the experiment was 9.74 x 10^{-5} m^3/DHF per m of icicle length which exceeds the estimation based on the Canadian experience. This growth rate was only observed once and therefore provides an encouraging but uncertain result. Consistent experimental growth rates of 3.25 x 10^{-5} m^3/DHF per m of icicle length were demonstrated which is 55% of the estimated achievable value. Due to the sub-optimal experimental conditions and, particularly, the high water supply temperature and uneven distribution to the icicle surface, the experimental results provide encouragement that the achievable growth rate based on the Canadian experience can be obtained.

The potential advantage of the icicle design over the Canadian ice block can be illustrated by the following comparison. Assume a full scale icicle production design uses 3 m long icicles and an icicle density of 4 icicles per m^2 of storage or land area. Using the 25,000 DHF available in the Canadian location and the achievable ice growth rate estimated above, the icicle system will produce 17.67 m^3 of ice per m^2 of land or storage area. This production is 2.4 times the 7.5 m^3/m^2 obtained by the Canadian ice block system.

4.2   Ice Production in U.S. Locations

The availability of freezing temperatures vary greatly in the U.S. as a function of latitude, ocean proximity, and other geographical conditions influencing climate. Using available typical meteorological year (TMY) hourly data from the National Climatic Center, the total DHF for locations in the Northeast U.S. and several other U.S. locations were calculated. The results are shown in Table 2. Typical values for northern locations appropriate for ice production are 12,000 to 18,000 DHF.

Combining the icicle productivity of 5.89 x 10^{-5} m^3/DHF per m of icicle length and the freezing availability of the northern U.S. of 15,000 DHF, the total ice growth anticipated from the enhanced icicle method is 0.884 m^3/m of icicle length. Assuming the design of 3 m icicles and 4 icicles per m^2 of land area, expected ice production in the U.S. climate would be 10.6 m^3/m^2 land area.

A useful perspective of this freezing rate can be gained by considering that during a one hour period of −10°C (14°F) temperature (10 DHF), 0.00177 m^3 of water would need to be frozen on a 3 m long icicle. The required flow rate of 0.00177 m^3/hr converts to 0.49 ml/sec (or about 4 drops per second of water supply). Provided the flow is well distributed around the icicle surface, this flow rate intuitively seems very reasonable in terms of its ability to freeze over the 3 m length of the icicle in −10°C weather. The flow rate at −5°C (23°F) would be half these values.

Table 2. Degree Hours of Freezing in U.S. Locations

U.S. LOCATION		DHF$_{°C}$
Burlington	VT	18,375
Bangor	ME	17,515
Madison	WI	17,154
Concord	NH	15,342
Binghamton	NY	14,898
Syracuse	NY	12,894
Albany	NY	12,494
Portland	ME	12,431
Rochester	NY	12,175
Buffalo	NY	10,836
Hartford	CT	9,688
Denver	CO	9,463
Pittsburgh	PA	9,135
Scranton	PA	8,520
Allentown	PA	7,378
Boston	MA	4,684

4.3 ENERGY AND ECONOMIC ASSESSMENT

4.3.1 Value of Ice as a Substitute for Conventional Cooling

Using the heat of fusion of water, 334 kJ/kg (144 BTU/lb), and the specific heat, 4.18 kJ/kg°K (1 BTU/lb°F), the useful cooling energy for building cooling from the ice and melt water up to 7°C (45°F) is 363 kJ/kg (157 BTU/lb). Using the density of ice, 1000 kg/m^3 (62.42 lb/ft^3), the energy value of the ice is 363 MJ/m^3 (9800 BTU/ft^3), or 101 kWh/m^3.

To determine the economic value of the ice used to displace conventional cooling, the following parameters and assumptions are assumed for the icicle system performance and for conventional air conditioning systems and electric utility rates.

Useful Energy Content of Ice	E_{ice}	101 kWh/m^3
Ice Storage Efficiency	η	0.8
Overall COP of Ice Storage System	COP_{ice}	25.0
Overall COP of Conventional A/C	$COP_{a/c}$	2.9
Purchased Electric Energy Cost (ice operation)	$C_{e,ice}$	0.05 \$/kWh
Displaced Electric Energy Cost (energy + demand savings from peak a/c)	$C_{e,a/c}$	0.12 \$/kWh

The following equation is used to determine the economic value of 1 m^3 of ice production. The last term reduces the cost savings due to displacing the conventional cooling by the operating cost for the ice system. The values used for electric costs assume an operational strategy for ice utilization which reduces peak summer electric demand charges. The analysis does not include any reduction in capital costs of the conventional system or substantial incentives often provided by utilities for peak kW reductions. Moreover, the costs do not include the substantial environmental and other social costs associated with conventional energy use.

$$\text{Economic Value of Ice (per m}^3) = (E_{ice})\,(\eta)\qquad\left(\frac{C_{e,a/c}}{COP_{a/c}} - \frac{C_{e,ice}}{COP_{ice}}\right)$$

$$= (101)\,(0.8)\,[(0.12/2.9) - (0.05/25)]$$
$$= 3.18\ \$/m^3$$

Electric and primary energy savings provided by a m^3 of ice produced by the icicle system and used to displace conventional cooling can be calculated by the following equations. Electric energy is assumed to be generated from primary fuel at a net efficiency of 0.30.

$$\text{Electric Energy Savings of Ice (per m}^3) = (E_{ice})(\eta)\qquad\left(\frac{1}{COP_{a/c}} - \frac{1}{COP_{ice}}\right)$$

$$= (101)\,(0.8)\,[(1/2.9) - (1/25)]$$
$$= 24.63\ kWh_e/m^3$$

$$\text{Primary Energy Savings of Ice (per m}^3)\ =\ \text{Electric Savings} / 0.30$$
$$= (24.63)/(0.30)$$
$$= 82.10\ kWh_{th}/m^3\,(296\ MJ/m^3,\ 280 \times 10^3\ BTU/m^3,\ 7933\ BTU/ft^3)$$

4.3.2 Cost Goals and Energy Savings for Icicle System

The value of ice can be used to determine cost goals for the ice storage system – that is, the maximum capital cost of the system which would make it cost effective. Although the cost of the icicle design has not been evaluated in detail, the cost goals may provide an initial sense of economic feasibility.

From Section 4.2, the height of solid ice which could be produced in a northern U.S. location was established to be 10.6 m. Combining this with the economic value of ice, the value of the ice produced in a year per m^2 of storage area is $33.73. Assuming a simple payback period of 10 years for the ice system capital investment (unlikely for private sector but appropriate for utility scale energy systems), the cost goal for project construction is $337/m^2 of storage area.

Economies of scale are substantial for the ice system – both for construction costs and thermal efficiency of the seasonal ice storage. Although the costs of the icicle enhancement surface, water distribution, and storage liner and insulation are relatively independent on project size, the pit excavation, load distribution piping, system enclosure, forced ventilation system, controls, and engineering exhibit large economies of scale.

A full scale system might have a base area of 50 x 50 m and annually produce 26,500 m³ of solid ice and serve the cooling load of a large commercial office building. Given the assumptions above, the system would annually displace \$84,319 of cooling payments and have a capital cost goal of \$843,190. This scale project would *annually* displace 653 MWh of electric energy or 7834 GJ (7.4 x 10⁹ BTU) of primary fuel – roughly equivalent to 1263 barrels of oil or 310 tons of coal.

The feasibility of the ice system meeting this cost goal is not adequately known; but the cost goal presents a large sum and in considered an encouraging result. The ice system is capable of displacing substantial quantities of conventional fuels and associated environmental and economic costs.

REFERENCES

1   Elliott, N. and Roberson, M-R., "Produce Cooling With Annual Ice Storage", *The Professional Engineer*, May–June 1990.

2   Kirkpatrick, D.L., Masoero, M., Rabl, A., Roedder, C.E., Socolow, R.H., and Taylor, T.B., "The Ice Pond – Production and Seasonal Storage of Ice for Cooling", The Center for Energy and Environmental Studies, Princeton University, PU/CEES Report No. 149, August 1982.

3   Kirkpatrick, D.L., Masoero, M., Rabl, A., Roedder, C.E., Socolow, R.H., and Taylor, T.B., "The Ice Pond – Production and Seasonal Storage of Ice for Cooling", *Solar Energy*, Vol. 35, No. 5, pp. 435–445, 1985.

4   Francis, C.E., "An Annual Storage Ice System for Summer Cooling", *Proceedings of ENERSTOCK '85, Third International Conference on Energy Storage for Building Heating and Cooling*, Toronto, Canada, September 1985.

5   Buies, S., "The *Fabrikaglace* Process to Make and Store Natural Ice", *Third International Workshop on Ice Storage for Cooling Applications*, Argonne National Laboratory, Illinois, ANL/CNSV–TM–177, March 1984.

6   "Enhanced Ice Pond Technology and Industrial Process Cooling: Final Report", New York State Energy Research and Development Authority, Agreement No. 1403–EEED–IE–90, August, 1990.

7   Poots, G., "On the Application of Integral-Methods to the Solution of Problems Involving the Solidification of Liquids Initially at Fusion Temperature", *International Journal of Heat and Mass Transfer*, Vol. 5, pp. 525–531, 1962.

8   Churchill, Stuart W., "Approximations for Conduction with Freezing or Melting", *International Journal of Heat and Mass Transfer*, Vol. 20, pp. 1251–1253, 1977.

9   Bell, Graham E., "Solidification of a Liquid About a Cylindrical Pipe", *International Journal of Heat and Mass Transfer*, Vol. 22, pp. 1681–1686, 1979.

10   Shih, Yen-Ping and Tsay, Sun-Yuan, "Analytical Solutions for Freezing a Saturated Liquid Inside or Outside Cylinders", *Chemical Engineering Science*, Vol. 26, pp. 809–816, 1971.

11   Cheng, K.C. and Sabhapathy, P., "Ice Formation Over an Isothermally Cooled Vertical Circular Cylinder in Natural Convection", *Journal of Heat Transfer*, Vol. 111, pp. 191–194, February 1989.

12   Cheng, K.C., Inaba, H., and Gilpin, R.R., "An Experimental Investigation of Ice Formation Around an Isothermally Cooled Cylinder in Crossflow", *Journal of Heat Transfer*, Vol. 103, pp. 733–738, November 1981.

13   Cheng, K.C., Inaba, H., and Gilpin, R.R., "Effects of Natural Convection on Ice Formation Around an Isothermally Cooled Horizontal Cylinder", *Journal of Heat Transfer*, Vol. 110, pp. 931–937, November 1988.

14   Lunardini, Virgil J., *Heat Transfer in Cold Climates*, Von Nostrand Reinhold Company, New York, 1981.

15   Cheng, K.C., Lunardini, V.J., and Seki, N., editors, *Proceedings of the 1987 International Symposium on Cold Regions Heat Transfer*, American Society of Mechanical Engineers, Alberta, Canada, June 1987.

16   Lozowski, E.P., Gates, E.M., and Makkonen, L., "Recent Progress in the Incorporation of Convective Heat Transfer into Cylindrical Ice Accretion Models", *Proceedings of the 1987 International Symposium on Cold Regions Heat Transfer*, American Society of Mechanical Engineers, Alberta, Canada, June 1987.

17   Lozowski, E.P., Stallabrass, J.R., and Hearty, P.F., "The Icing of an Unheated, Nonrotating Cylinder. Part 1: A Simulation Model", *Journal of Climate and Applied Meteorology*, Vol. 22, pp. 2053–2062, December 1983.

Modeling the Direct Microbial Conversion of Lignocellulose to Ethanol by *Clostridium thermocellum* in an Upflow Solids Retaining Bioreactor

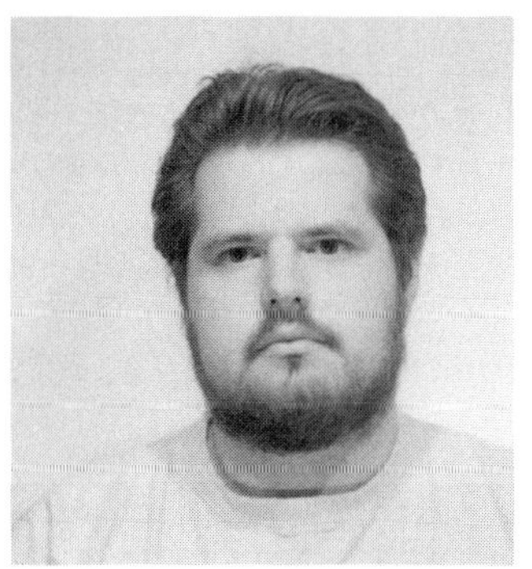

DAVID ANTHONY HOGSETT

Thayer School of Engineering
Dartmouth College
Research Supervisor: Dr Lee R. Lynd

ABSTRACT

An Upflow Solids Retaining Bioreactor (USRB) is proposed for the bioconversion of lignocellulose to ethanol using the thermophilic bacterium *Clostridium thermocellum*. Modeling of the USRB indicates the design may significantly reduce the large bioconversion costs in present methods of cellulose hydrolysis and fermentation. Potential advantages of the reactor concept include retention and beneficial stratification of substrate and biocatalysts. Such retention and stratification, if achieved, should improve cellulose conversion and reactor throughput. A chemostat model for direct microbial conversion of lignocellulose is developed and used to evaluate predicted performance for two simple model conceptions of the USRB. At high substrate conversions necessary for economic processes, the USRB is predicted to exhibit a tenfold dilution advantage over the chemostat.

OVERVIEW

The United States annually consumes 15 quads of gasoline to meet the transportation needs of its 180 million vehicle fleet.[15] Domestic crude oil production covers just half of U.S. petroleum consumption, the remainder coming from a variety of foreign reserves.[15] Foreign oil importation accounts for over one-half of the U.S. trade deficit and repeatedly results in international price and supply instability. Beyond supply issues, domestic combustion of petroleum generates substantial air pollution including 900 million tons of carbon dioxide.[53] The fall of ambient air quality below federal ozone guidelines for nearly one-half of all Americans is traceable, in part, to current gasoline usage patterns.[53] Conservation and new combustion technology may reduce pollution, carbon dioxide emissions, and the need for imported oil but a different, gasoline-displacing fuel is likely necessary to eliminate the problems.[53,37]

Ethanol produced from biomass is a tenable liquid fuel option to gasoline which can both meet fuel demands and relieve the intractable problems of gasoline.[37] Ethanol can be both blended with gasoline, as it is in 7% of the United States gasoline, or combusted neatly, as presently done in Brazil.[26] Beyond positive impacts on trade and the environment, ethanol compares favorably to gasoline and other alternative liquid fuels as a transportation fuel. Most problems traditionally associated with ethanol powered vehicles have been overcome and ethanol-optimized engine designs are available.[37]

There are many logistical, economic, and technological barriers to widespread utilization of fuel ethanol in the United States. Unlike Brazil, the U.S. cannot rely upon massive, subsidized sugar crops; consequently, corn is the domestic feedstock of choice.[64] While corn has many favorable features as a process feedstock, it has limited gross availability, trades as a food commodity, and its fuel utilization does not have a large energy return.[13,37]

Biomass, specifically cellulosic plants, has the potential to replace corn as the predominant industrial feedstock and allow for a large expansion of the ethanol fuel market.[37] Projected to be available in quantities sufficient to meet most domestic transportation fuel needs, cellulosics are a relatively cheap feedstock with a high energy return.[37,21] Unfortunately, cellulosics are inherently more difficult and currently more expensive to convert to ethanol than are either raw sugar or corn.[14,61,64] In current processes for converting cellulosics to ethanol, 30% of the production cost is due to the biologically mediated steps of enzymatic hydrolysis and cellular fermentation.[64] In addition to their expense, these bioconversions are the least developed process steps and potentially most rewarding areas for process improvement.[64,37,23,40,61]

Direct microbial conversion (DMC), the single vessel hydrolysis and fermentation of cellulose to ethanol, offers the potential for dramatic cost reductions, on the order of 25%, over standard approaches.[23] DMC has been proposed for several microbial systems, the most studied system utilizing a bacterial coculture of the cellulase producing, hexose fermenting *Clostridium thermocellum*, and the pentose fermenting *Clostridium thermosaccharolyticum*.[4,30] To date, progress in Clostridia based DMC has been retarded by inadequate fermentation performance, indicated by poor ethanol yields and reportedly low microbial ethanol tolerance.[36,22] However, it is hydrolysis, not fermentation, which limits the rate of the overall process of converting cellulose to ethanol and it is DMC's comparatively rapid, inexpensive hydrolysis which distinguishes the approach.

An Upflow Solids Retaining Bioreactor (USRB) exploits the key advantage of DMC by addressing the limitation of hydrolysis. In the USRB, Figure 1, a cellulosic slurry is introduced at the top of the reactor and allowed to gravity settle. Solids are removed from the bottom at a rate necessary to maintain bed height. Liquid removed from above the settled bed is continuously recycled up through the bed to provide acceptable liquid phase composition.

It is hypothesized that under continuous cultivation in the USRB the largely quiescent bed of cellulose and lignin will act to concentrate both enzymatic and microbial catalysts, allowing for increased reactor throughput and higher levels of substrate utilization than is attainable in the typical continuous stirred tank reactor (CSTR) configuration. Additional advantages of the USRB may include higher product yields in a bioreactor design which is both robust and suitable for continuous product recovery. The USRB may potentially have an additional role in processes which do not utilize DMC, such as the Simultaneous Saccharification and Fermentation processes under investigation by other researchers.

BACKGROUND

Lignocellulose

Lignocellulose is perhaps the most abundant renewable natural resource available to mankind. Composed primarily of cellulose, hemicellulose, and lignin, lignocellulose is potentially available in quantities sufficient to meet transportation demands for motor fuel.[37] Each cellulose molecule is a long, unbranched chain of d-glucose subunits with both amorphous and crystalline structural characteristics in native sources. Hemicelluloses are shorter, branched pentose and hexose polymers. Lignin is a polyphenoic compound which resists both chemical and enzymatic attack. In most sources of woody lignocellulose cellulose, hemicellulose, and lignin are in approximate mass ratios of 1.5:1:1.[5]

Cellulose is naturally resistant to process degradation and requires some method of pretreatment to permit high substrate conversions via enzymatic hydrolysis. Many techniques have been proposed including steam explosion, dilute acid hydrolysis, ammonia freeze explosion, and others. Technology for lignocellulose pretreatment is immature and will likely change before an economic process for ethanol production from lignocellulose emerges.[61] Preliminary continuous culture work with *Clostridium thermocellum* has utilized dilute sulfuric acid steam pretreated hardwoods.[39]

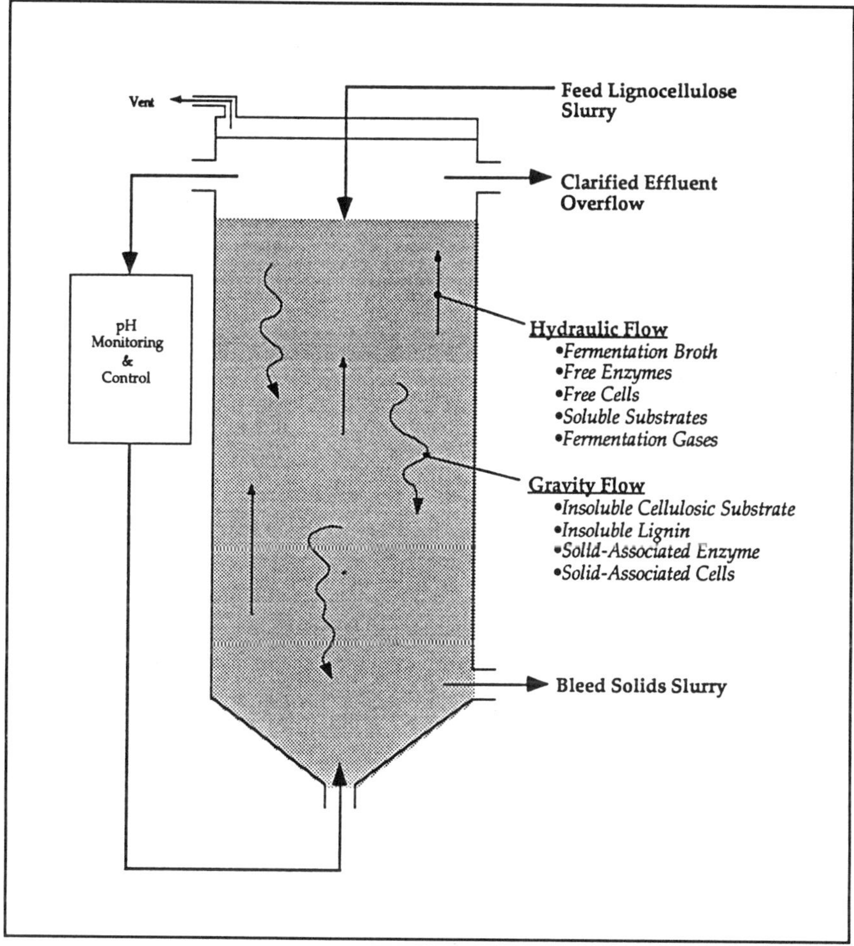

Figure 1. Overview of design and operation of an Upflow Solids Retaining Bioreactor showing anticipated flow phenomena. Bleed rate is controlled so as to maintain the largest possible settled bed. Recycle rates are set by the operator within the confines of the bed's fluidization behavior.

Clostridium thermocellum

Clostridium thermocellum is an anaerobic thermophile with optimum growth temperature of 60°C. Possessing a gram positive-type cell membrane, the organism stains gram variable. Growth of the organism occurs on both cellulose and cellobiose with some strains apparently possessing the ability to utilize glucose as well. Major end products of fermentation are acetate, ethanol, lactate, carbon dioxide, and hydrogen gas.[4,39] Most strains produce cellulase and adhere to cellulose at some developmental stage, although cellulase production is repressed by cellobiose and non-cellulase producing and/or non-binding mutant strains have been developed.[7]

The distinguishing feature of *C. thermocellum* in an applied context is its cellulase complex, or cellulosome.[35,23] The cellulosome from *Clostridium thermocellum* is a multicomponent complex which rapidly hydrolyzes crystalline cellulose, producing almost exclusively cellobiose as an end product. With a molecular weight of 2 to 6 million, the cellulosome's activity is inhibited by cellobiose but not glucose, and is stabilized by calcium ions and thiol reagents.[29,24,33] Understanding of the cellulosome has been retarded by the difficulty of separating the recalcitrant complex into its component parts.[52] Subcellulosome preparations composed of six main protein subunits (molecular weight 58,000 to 210,000) have been shown to degrade crystalline cellulose at rates 8-fold higher than the crude extracellular cellulosome.[29] Combined preparations of cellulosome and β-glucosidase from *A. niger* have been shown to completely hydrolyze concentrated Avicel suspensions.[33]

Endoglucanases, xylanases, and β-glucosidases have been purified and characterized and genes for each enzyme have been reported.[32 43] Genes coding for endoglucanases have been cloned and expressed in *E. coli*. Recently, a major cellobiohydrolase, which requires calcium ions for stability, has been isolated from the complex.[45] β-glucosidases appear to be almost exclusively located outside of the cellulosome on the cell surface, with none detected in extracellular preparations.[9] The cellulosome also appears to be located primarily on the cell surface, organized into protuberances which contain hundreds of cellulosomes.[32] During early exponential growth the cellulosome appears to be tightly packed, with increasing looseness accompanying cell maturity.[41] Dramatic changes of the cellulosome packed protuberances appear to occur with direct cellulose contact.[8,7] Adherence of *C. thermocellum* to cellulosics has been attributed to the surface organelles and has been used for isolation and purification of strains.[7,62] Preliminary data on adherence of *C. thermocellum* to cellulose has been established.[28] Quantification of adherence may be desirable in the context of this research and can likely be achieved using general techniques.[51,17]

High affinity and rapid adsorption of the crude extracellular cellulosome to Avicel, lignin, and diluted acid pretreated hardwoods has been shown.[10] Hydrolysis kinetics have been experimentally determined and modeled using modified Michelis-Menten techniques.[10,1] *In vitro* hydrolysis rates and conversions for pretreated hardwoods appear to be similar for *C. thermocellum* and *T. reesei* cellulase at approximately equal loadings, with complete hydrolysis in under 24 hrs.[38,39] In contrast to other cellulases, the degradation of amorphous and crystalline cellulose by the cellulosome apparently occurs at equal rates.[49]

Clostridium thermocellum utilizes substrate level phosphorylation for energy generation. The mixed solvent formation, typically ethanol, acetate, and lactate mass ratios of 10:10:1 in continuous culture, are the primary limitation of the organism for ethanol production.[39,36,4] While improving ethanol yield is not a specific design goal of the USRB, higher yields have been associated with elevated dissolved hydrogen concentrations which may be anticipated in an USRB with low agitation.[34,31] Improvement of ethanol yield may also be feasible via genetic or metabolic approaches.[30] Improving yields is an prime research area at the Thayer School of Engineering with work being primarily conducted using the pentose fermenting thermophile, *Clostridium thermosaccharolyticum*. Low ethanol tolerance has been proposed as a second major drawback of thermophilic fermenters but recent work suggest that thermophiles are more robust in the presence of elevated ethanol concentrations when grown in continuous culture.[22,36]

Direct Microbial Conversion

In the bioproduction of ethanol from lignocellulose, four biologically-mediated steps can be identified: production of cellulase enzymes, hydrolysis of insoluble substrates, hexose fermentation, and pentose fermentation. Direct microbial conversion (DMC) processes achieve enzyme production, substrate hydrolysis, and fermentation to ethanol in a single process step and can be contrasted to processes which separate the biological steps. DMC processes have been proposed using fungi, bacteria, and yeast and topics relevant to DMC have been pursued by several laboratories in the last decade and are the subject of multiple reviews.[23] The possible advantage of DMC lies in its reduced bioconversion costs and potential for higher overall yields.[23] The simplicity of DMC is also compatible with producing a low-value commodity and may allow rapid process development if key technical issues can be resolved.

The primary incentive for pursuing DMC is the dramatic impact of cellulase production costs on overall process economics. Cellulase production in DMC systems may involve very small or zero cost making cellulase essentially free. The economic consequences of free, plentiful cellulase are pervasive and have the potential of dramatically cutting overall costs of ethanol production from lignocellulose.[23] Other advantages of specific DMC processes have been reported, including compatible pentose and hexose fermenters and cellulase, higher kinetic rates, reduced cooling costs, and reduced distillation expense.[23]

Differential Retention and Bioreactor Design

Development of high productivity reactors for fermentation has focused primarily on retaining cells, and to a lesser extent product removal.[50,44,42,6,63] Cell retaining bioreactors have been described, modeled, and are widely used in the production of both high and low value products.[25,60,20] Though very effective for the conversion of soluble substrates, most cell retaining reactors rely on diffusion driven substrate delivery and are unsuited for insoluble substrates. Recycling of cells in the presence of an inert insoluble fraction, as would be necessary in the case of lignocellulose conversion, is also both expensive and potentially counterproductive.

While concentrating or retaining soluble substrates is often problematic, the retention of insoluble substrates is relatively straightforward and many designs have been proposed for waste water treatment, methane production, and coal desulfurization.[54,55,56,27,19,16,3] Columnar reactors for cellulose hydrolysis have been described, several of which utilize cellulase recycling via filtration.[57,58,18] Reactor schemes which combine both columnar cellulose hydrolysis and cell retaining fermentation have also been investigated.[47,59] In all cases known to the author, cellulose loading of the hydrolysis or hydrolysis/fermentation systems has been discontinuous and required reaction vessels separate from the fermenter.

The hydrodynamic characteristics of pretreated lignocellulosics has been limited to fluidization studies of non-reacting Aspen chips.[11,12] The influence of cell adhesion and flocculation has been given attention, but only for fluidized beds with non-reacting growth supports.[46,48] To the author's knowledge, hindered settling of a reacting substrate with attached microbial growth in the presence of a sub-fluidizing continuous phase has not been considered in detail for any insoluble substrate.

ANALYSIS

The Upflow Solids-Retaining Bioreactor (USRB) is a single stage anaerobic reactor for aqueous fermentation of solid substrates. In its present configuration for lignocellulosic fermentation to solvents by *Clostridium thermocellum*, a slurry of pretreated lignocellulose is continuously fed into the USRB at a point below the effluent and recycle ports. During preliminary experimentation at low recycle rates, an USRB develops two distinct regimes; solids in the feed slurry settle towards the bottom of the reactor and form a bed and a clarified zone is established above the feed port.

In operating an USRB, a bleed flow is chosen so as to maintain the largest bed volume possible without significant solids washout. Liquid from the clarified regime is recycled at a rate adequate to assure proper pH levels. After pH adjustment, the solid free fluid is pumped into the reactor base. Fermentation gases are freely vented from the reactor headspace. Temperature control is achieved through reactor jacketing.

Possible advantages of an USRB include increased solid and biocatalyst residence times at high dilution rates and the development of stratified substrate and catalyst profiles. Modeling the continuous hydrolysis and fermentation of lignocellulose in a CSTR provides a foil against which the potential advantages of an USRB can be assessed.

Continuous Stirred Tank Reactor (CSTR)

A conceptual picture of a CSTR for continuous culture is shown in Table 1 with model nomenclature, parameters, and assumptions detailed in Tables 4–6. For the CSTR shown, two phase mixtures flow in and out of the reactor. The exiting stream has a composition identical to the reactor itself. In such a case, the equations shown can be used to describe the condition of state variables. At steady state, the time differentials are equal to zero, and the system of equations has a unique solution. The derivation and numerical investigation of the CSTR model requires making several assumptions, listed below.

CSTR Model Assumptions

- Solids and liquids are ideally mixed in the fermenter
- Gaseous products and voids are ignored
- Cell growth follows Monod kinetics
- Linear ethanol inhibition of cell growth
- Michelis-Menten type cellulase adsorption
- Michelis-Menten type hydrolysis rates
- Constant yields of cells and ethanol
- Intrinsic parameters from a variety of sources
- Operation at 5% cellulose feed with 3.5% inert lignin

Table 1. CSTR Modeling Equations

1. $$\frac{dC}{dt} = \left(\frac{Q_i\,\varepsilon_i}{V\,\varepsilon_n}\right) C_i - \left(\frac{Q_o}{V}\right) C - (k)\,EC$$

2. $$\frac{dL}{dt} = \left(\frac{Q_i\,\varepsilon_i}{V\,\varepsilon_o}\right) L_i - \left(\frac{Q_o}{V}\right) L$$

3. $$\frac{dX}{dt} = (\mu)\,X - \left(\frac{Q_o}{V}\right) X$$

4. $$\frac{dE}{dt} = (\mu Y_{EX})\,X - \left(\frac{Q_o}{V}\right) E$$

5. $$\frac{dG}{dt} = (k)\,EC - \left(\frac{Q_o}{V}\right) G - \left(\frac{\mu}{Y_{XS}}\right) X$$

6. $$\frac{dP}{dt} = (\mu Y_{PX})\,X - \left(\frac{Q_o}{V}\right) P$$

7. $$Q_i = Q_o$$

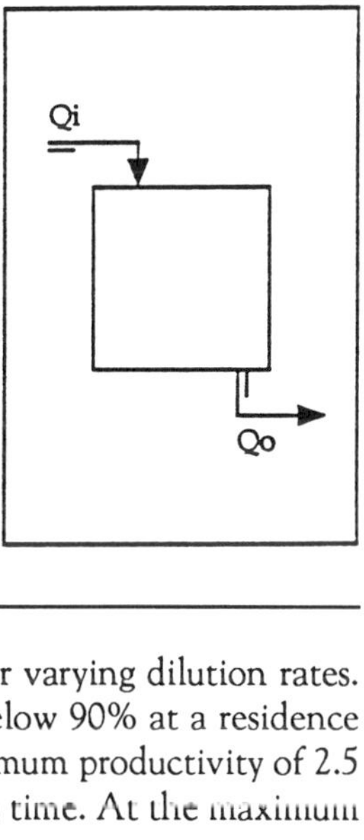

Figure 2 shows the performance of the CSTR under varying dilution rates. Modeling predicts that substrate utilization will fall below 90% at a residence time of approximately 12 hours (D = .08) with a maximum productivity of 2.5 g/l/hr at a dilution corresponding to a 7 hour residence time. At the maximum productivity, substrate utilization has fallen to 75%.

Upflow Solids Retaining Bioreactor (USRB)

Two conceptual pictures of an USRB for continuous culture are shown in Tables 2 and 3. Consideration of the one bed USRB model permits isolation of the impact of solids concentration and retention without the added complexity of stratification. The two bed USRB model builds upon the single bed and simulates the combined impact of retention and stratification.

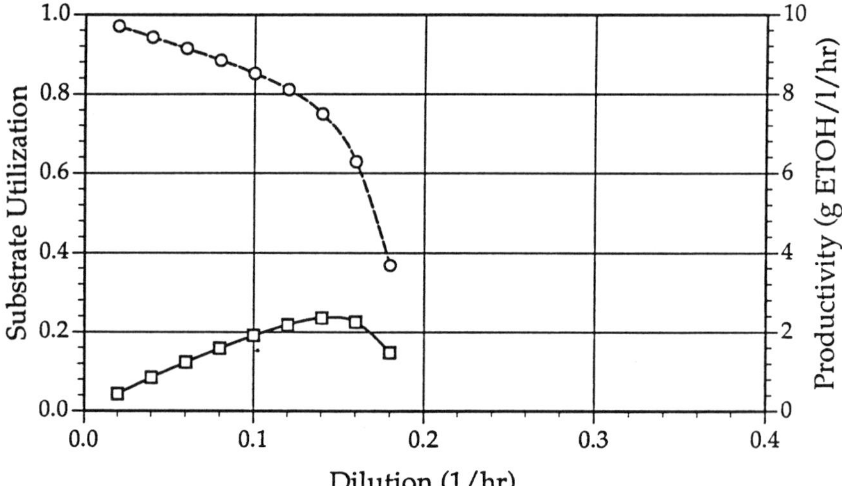

Figure 2. Model prediction of utilization (O) and productivity (□) of a CSTR for direct microbial conversion of lignocellulose to ethanol using C. *thermocellum.*

For the one bed USRB conceptualized, Table 2, a single well mixed bed contains a concentration of the solid substrates and adsorbed biocatalysts. A two phase flow exits the reactor base with a composition identical to that of the concentrated bed. At the reactor top, a single phase flow leaves with composition equal to the solid-free zone. Conservation of lignin provides basis for establishing the exiting volumetric flows. In such a case, the equations shown can be used to describe the condition of state variables. At steady state, the time differentials are equal to zero, and the system of equations has a unique solution. The derivation and numerical investigation of the USRB models requires making several assumptions, listed below.

USRB Model Assumptions

- Solids and liquids are ideally mixed in the fermenter sections
- Gaseous products and voids are ignored
- Cell growth follows Monod kinetics
- Linear ethanol inhibition of cell growth
- Michelis-Menten type cellulase adsorption and hydrolysis
- Constant yields of cells, cellulase, and ethanol
- Intrinsic parameters from a variety of sources
- Operation at 5% cellulose feed with 3.5% inert lignin
- Recycle flow is minimally 4 times higher than feed flow

Table 2. USRB Modeling Equations – One Well Mixed Bed

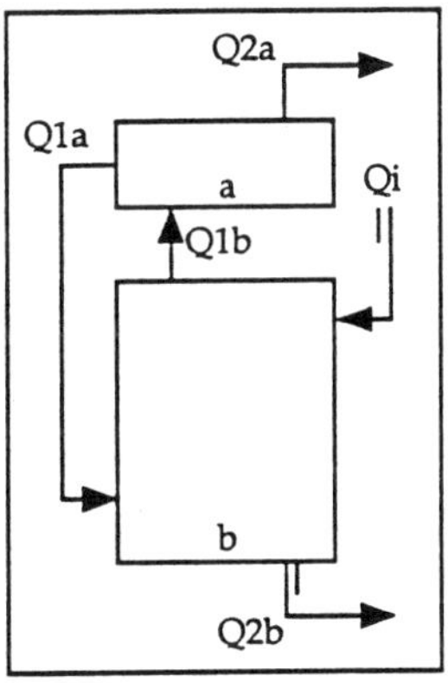

1. $\dfrac{dX_a}{dt} = (\mu_a) X_a + \left(\dfrac{Q_{1b}}{V_a}\right) Xf_b - \left(\dfrac{Q_{1a} + Q_{2a}}{V_a}\right) X_a$

2. $\dfrac{dE_a}{dt} = (\mu_a Y_{EX_a}) X_a + \left(\dfrac{Q_{1b}}{V_a}\right) Ef_b - \left(\dfrac{Q_{1a} + Q_{2a}}{V_a}\right) E_a - (k_1) E_a$

3. $\dfrac{dG_a}{dt} = \left(\dfrac{Q_{1b}}{V_a}\right) G_b - \left(\dfrac{Q_{1a} + Q_{2a}}{V_a}\right) G_a - \left(\dfrac{\mu_a}{Y_{XS}}\right) X_a$

4. $\dfrac{dP_a}{dt} = (\mu_a Y_{PX}) X_a + \left(\dfrac{Q_{1b}}{V_a}\right) P_b - \left(\dfrac{Q_{1a} + Q_{2a}}{V_a}\right) P_a$

5. $\dfrac{dC_b}{dt} = \left(\dfrac{Q_i \varepsilon_i}{V_b \varepsilon_b}\right) C_i - \left(\dfrac{Q_{2b}}{V_b}\right) C_b - (k_b) E C_b$

6. $\dfrac{dX_b}{dt} = \left(\dfrac{Q_{1a}}{V_b \varepsilon_b}\right) X_a + (\mu_b) X_b - \left(\dfrac{Q_{1b}}{V_b \varepsilon_b}\right) Xf_b - \left(\dfrac{Q_{2b}}{V_b}\right) X_b$

7. $\dfrac{dE_b}{dt} = \left(\dfrac{Q_{1a}}{V_b \varepsilon_b}\right) E_a + (\mu_b Y_{EX_b}) X_b - \left(\dfrac{Q_{1b}}{V_b \varepsilon_b}\right) Ef_b - \left(\dfrac{Q_{2b}}{V_b}\right) E_b - (k_1) E_b$

8. $\dfrac{dG_b}{dt} = \left(\dfrac{Q_{1a}}{V_b \varepsilon_b}\right) G_a + (k_b) E C_b - \left(\dfrac{\mu_b}{Y_{XS}}\right) X_b - \left(\dfrac{Q_{1b}}{V_b \varepsilon_b}\right) G_b - \left(\dfrac{Q_{2b}}{V_b}\right) G_b$

9. $\dfrac{dP_b}{dt} = \left(\dfrac{Q_{1a}}{V_b \varepsilon_b}\right) P_a + (\mu_b Y_{PX}) X_b - \left(\dfrac{Q_{1b}}{V_b \varepsilon_b}\right) P_b - \left(\dfrac{Q_{2b}}{V_b}\right) P_b$

10. $Q_{2b} = \left(\dfrac{1 - \varepsilon_i}{1 - \varepsilon_b}\right)\left(\dfrac{1 - Fc_i}{1 - Fc_b}\right) Q_i$

11. $Q_{2a} = Q_i - Q_{2b}$

12. $Q_{1b} = Q_{2a} + Q_{1a}$

Figure 3 shows the performance of the one bed USRB model conception under varying dilution rates. Modeling predicts that substrate utilization will fall below 90% at a residence time of approximately 5 hours (D = .20) with a maximum productivity of nearly 7 g/l/hr at a dilution corresponding to a 3 hour residence time. At the maximum productivity, substrate utilization has fallen to 85%.

For the two bed USRB conceptualized, Table 3, two well mixed beds contain a concentration of the solid substrates and adsorbed biocatalysts. A two phase flow exits the reactor base with a composition identical to that of the concentrated bed. At the reactor top, a single phase flow leaves with composition equal to the solid-free zone. In this conception, single phases flow within the bed volumes. Conservation of lignin provides basis for establishing internal and exiting volumetric flows. In such a case, the equations shown can be used to describe the condition of state variables. At steady state, the time differentials are equal to zero, and the system of equations has a unique solution.

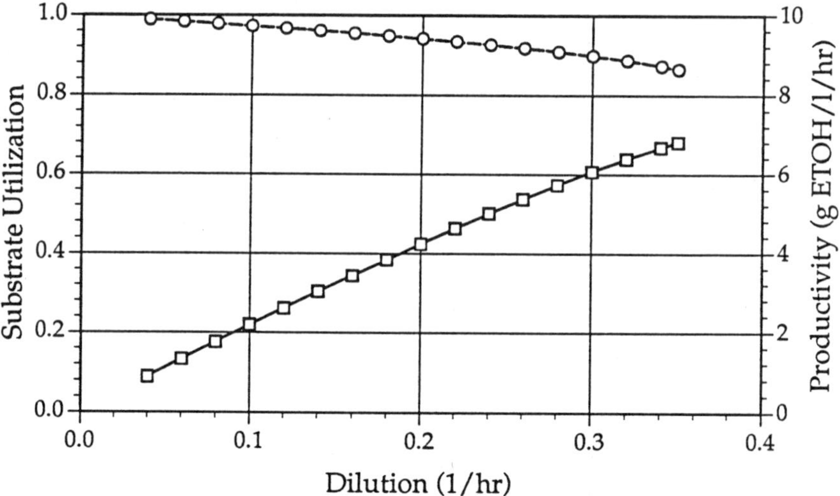

Figure 3. One bed model prediction of utilization (O) and productivity (□) of an USRB for direct microbial conversion of lignocellulose to ethanol using C. *thermocellum.*

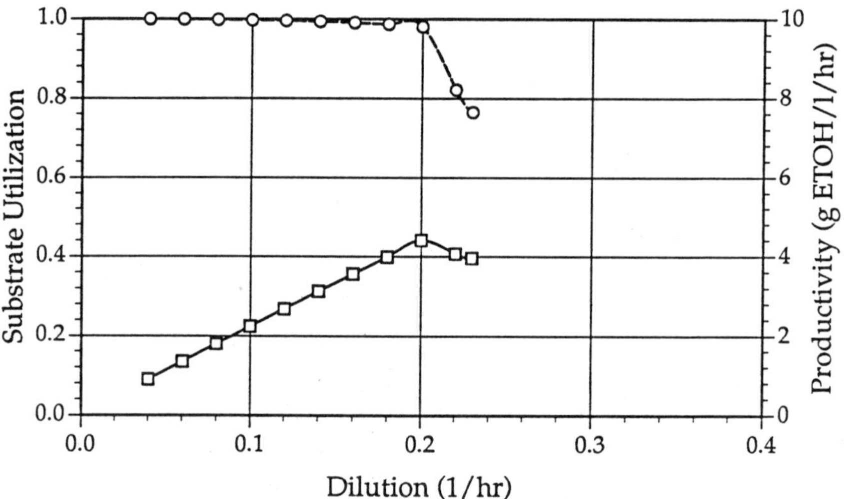

Figure 4. Two bed model prediction of utilization (O) and productivity (□) of an USRB for direct microbial conversion of lignocellulose to ethanol using C. *thermocellum.*

Table 3. USRB Modeling Equations – Two Well Mixed Beds

1. $$\frac{dC_b}{dt} = \left(\frac{Q_i \varepsilon_i}{V_b \varepsilon_b}\right) C_i - \left(\frac{\rho F c_b Q_{2b}}{V_b \varepsilon_b}\right) - (k) E C_b$$

2. $$\frac{dC_c}{dt} = \left(\frac{F c_b Q_{2b}}{\rho V_c \varepsilon_c}\right) - (k) E C_c - \left(\frac{Q_{2c}}{V_c}\right) C_c$$

3. $$\frac{dG_a}{dt} = \left(\frac{Q_{1b}}{V_a}\right) G_b - \left(\frac{Q_{1a} + Q_{2a}}{V_a}\right) G_a - \left(\frac{\mu_a}{Y_{XS}}\right) X_a$$

4. $$\frac{dG_b}{dt} = \left(\frac{Q_{1c}}{V_b \varepsilon_b}\right) G_c + (k) E C_b - \left(\frac{\mu_b}{Y_{XS}}\right) X_b - \left(\frac{Q_{1b}}{V_b \varepsilon_b}\right) G_b$$

5. $$\frac{dG_c}{dt} = \left(\frac{Q_{1a}}{V_c \varepsilon_c}\right) G_a + (k) E C_c - \left(\frac{\mu_c}{Y_{XS}}\right) X_c - \left(\frac{Q_{1c}}{V_c \varepsilon_c}\right) G_c - \left(\frac{Q_{2c}}{V_c}\right) G_c$$

6. $$\frac{dP_a}{dt} = (\mu_a Y_{PX}) X_a + \left(\frac{Q_{1b}}{V_a}\right) P_b - \left(\frac{Q_{1a} + Q_{2a}}{V_a}\right) P_a$$

7. $$\frac{dP_b}{dt} = \left(\frac{Q_{1c}}{V_b \varepsilon_b}\right) P_c + (\mu_b Y_{PX}) X_b - \left(\frac{Q_{1b}}{V_b \varepsilon_b}\right) P_b$$

8. $$\frac{dP_c}{dt} = \left(\frac{Q_{1a}}{V_c \varepsilon_c}\right) P_a + (\mu_c Y_{PX}) X_c - \left(\frac{Q_{1c}}{V_c \varepsilon_c}\right) P_c - \left(\frac{Q_{2c}}{V_c}\right) P_c$$

9. $$\frac{dX_a}{dt} = (\mu_a) X_a + \left(\frac{Q_{1b}}{V_a}\right) X f_b - \left(\frac{Q_{1a} + Q_{2a}}{V_a}\right) X_a$$

10. $$\frac{dX_b}{dt} = \left(\frac{Q_{1c}}{V_b \varepsilon_b}\right) X f_c + (\mu_b) X_b - \left(\frac{Q_{1b}}{V_b \varepsilon_b}\right) X f_b - \left(\frac{\rho F c_b Q_{2b}}{V_b \varepsilon_b}\right) \frac{X C_b}{C_b} - \left(\frac{\rho (1 - F c_b) Q_{2b}}{V_b \varepsilon_b}\right) \frac{X L_b}{L_b}$$

11. $$\frac{dX_c}{dt} = \left(\frac{Q_{1a}}{V_c \varepsilon_c}\right) X_a + (\mu_c) X_c + \left(\frac{\rho F c_b Q_{2b}}{V_c \varepsilon_c}\right) \frac{X C_b}{C_b} + \left(\frac{\rho (1 - F c_b) Q_{2b}}{V_c \varepsilon_c}\right) \frac{X L_b}{L_b} - \left(\frac{Q_{1c}}{V_c \varepsilon_c}\right) X f_c - \left(\frac{Q_{2c}}{V_c}\right) X_c$$

12. $$\frac{dE_a}{dt} = (\mu_a Y_{EX}) X_a + \left(\frac{Q_{1b}}{V_a}\right) E f_b - \left(\frac{Q_{1a} + Q_{2a}}{V_a}\right) E_a$$

13. $$\frac{dE_b}{dt} = \left(\frac{Q_{1c}}{V_b \varepsilon_b}\right) E_c + (\mu_b Y_{EX}) X_b - \left(\frac{Q_{1b}}{V_b \varepsilon_b}\right) E f_b - \left(\frac{\rho F c_b Q_{2b}}{V_b \varepsilon_b}\right) \frac{E C_b}{C_b} - \left(\frac{\rho (1 - F c_b) Q_{2b}}{V_b \varepsilon_b}\right) \frac{E L_b}{L_b}$$

14. $$\frac{dE_c}{dt} = \left(\frac{Q_{1a}}{V_c \varepsilon_c}\right) E_a + (\mu_c Y_{EX}) X_c + \left(\frac{\rho F c_b Q_{2b}}{V_c \varepsilon_c}\right) \frac{E C_b}{C_b} + \left(\frac{\rho (1 - F c_b) Q_{2b}}{V_c \varepsilon_c}\right) \frac{E L_b}{L_b} - \left(\frac{Q_{1c}}{V_c \varepsilon_c}\right) E f_c - \left(\frac{Q_{2c}}{V_c}\right) E_c$$

Figure 4 shows the performance of the two bed USRB model conception under varying dilution rates. Modeling predicts that substrate utilization will fall below 90% at a residence time of approximately 5 hours (D = .20) with a maximum productivity of 4.5 g/l/hr at the same dilution rate.

Table 4. Model Nomenclature

State Variables

C	Cellulose	grams cellulose/liquid liter
E	Cellulase	units activity/liquid liter
G	Soluble Sugar	grams soluble sugar/liquid liter
L	Lignin	grams lignin/liquid liter
P	Ethanol	grams ethanol/liquid liter
X	Cells	grams dry cell weight/liquid liter

Organism Characteristics

μ	Growth Rate	reciprocal hours
μ_{max}	Growth Rate Max.	reciprocal hours
K_G	Monod Saturation	grams soluble sugar/liquid liter
P_{max}	Ethanol Inhib. Const.	grams ethanol/liquid liter
Y_{EXmax}	Cellulase Yield	units activity/gram dry cell mass
G_{max}	Synthesis Repression	grams soluble sugar/liquid liter
K_{EC}	E <–> C Adsorp. Equil.	liquid liters/gram cellulose
K_{EL}	E <–> C Adsorp. Equil.	liquid liters/gram lignin
K_{XC}	E <–> C Adsorp. Equil.	liquid liters/gram cellulose
K_{XL}	E <–> C Adsorp. Equil.	liquid liters/gram lignin
B_{EC}	Binding Capacity	units activity/gram cellulose
B_{EL}	Binding Capacity	units activity/gram lignin
B_{XC}	Binding Capacity	grams dry cell weight/gram cellulose
B_{XL}	Binding Capacity	grams dry cell weight/gram lignin
k_{max}	Hydrolysis Rate Max.	grams hydrolyzed/unit complexed/hour
P_{2max}	Ethanol Inhib. Const.	grams ethanol/liquid liter
K_{G1}	Soluble Sugar Inhib.	grams soluble sugar/liquid liter
Y_{XS}	Cell Yield	gram dry cell weight/gram soluble sugar
Y_{PX}	Ethanol Yield	grams ethanol/gram dry cell mass

Volumetric Characteristics

Q	Volumetric Flow	liters/hour
V	Volume	liters
ε	Void Fraction	liter liquid/liter total
Fc	Fraction Cellulose	grams cellulose/gram solid
ρ	Solid Density	grams solid/liter solid

Table 5. Evaluation of Parameters

Parameter	Value	Ref	Comments
μ_{max}	.44	[2]	reported for *C. thermosaccharolyticum*
K_G	.11	[2]	reported for *C. thermosaccharolyticum*
P_{max}	52.6	[2]	reported for *C. thermosaccharolyticum*
Y_{EXmax}	3750	[10,39]	calculated using values from both sources
G_{max}	2	[47]	representative value
K_{EC}	335.9	[10]	dilute acid pretreated hardwood
K_{EL}	16.4	[10]	dilute acid pretreated hardwood
K_{XC}	.02		heuristic
K_{XL}	.02		heuristic
B_{EC}	1526	[10]	dilute acid pretreated hardwood
B_{EL}	1141	[10]	dilute acid pretreated hardwood
B_{XC}	.1		heuristic
B_{XL}	.1		heuristic

Parameter	Value	Ref	Comments
k_{max}	.00057	10,39	calculated using values from both sources
P_{2max}	160	10	measured in absence of substrate
K_{G1}	1.5	33	numerical fit of data
Y_{XS}	.11	23	representative of similar organisms
Y_{PX}	4.1	23	representative of similar organisms
ρ	1360		dilute acid pretreated hardwood
C_i	53.2		assumed operational condition
L_i	33.5		assumed operational condition

Table 6. Fundamental Modeling Equations

Growth of the organism follows Monod kinetics but is inhibited by ethanol; cellulase production is repressed by soluble sugars.

$$\mu = \left(\frac{\mu_{MAX}G}{K_G + G}\right)\left(1 - \frac{P}{P_{MAX}}\right) \qquad Y_{EX} = Y_{EXmax}\left(1 - \frac{G_i}{G_{max}}\right)$$

Michelis-Menten type enzyme adsorption to cellulose and lignin with noncompetitive adsorption of cells to cellulose and lignin to different sites.

$$K_{EC} = \frac{(EC)}{(Ef)(Cf)} \qquad K_{EL} = \frac{(EL)}{(Ef)(Lf)}$$

$$K_{XC} = \frac{(XC)}{(Xf)(Cf)} \qquad K_{XL} = \frac{(XL)}{(Xf)(Lf)}$$

Enzymes and cells can either be free or adsorbed to an insoluble fraction with cellulose and lignin having a limited capacity to bind either biocatalyst.

$$E = (Ef) + (EC) + (EL) \qquad X = (Xf) + (XC) + (XL)$$

$$C = Cf_E + \frac{EC}{B_{EC}} = Cf_X + \frac{XC}{B_{XC}} \qquad L = Lf_E + \frac{EL}{B_{EL}} = Lf_X + \frac{XL}{B_{XL}}$$

Hydrolysis rate is proportional to enzyme adsorbed to cellulose with hydrolysis inhibited by both ethanol and soluble sugar.

$$Rate_i = k_i(EC_i) \qquad k_i = k_{max}\left(1 - \frac{P_i}{P_{2max}}\right)\left(\frac{K_{G1}}{K_{G1} + G_i}\right)$$

In the USRB, settled insoluble substrate voidage is described by a simple function of the fraction of cellulose in the bed.

$$\varepsilon_i = 0.1Fc_i + 0.85 \qquad Fc_i = \frac{C_i}{C_i + L_i}$$

Discussion of Simulation Results

The predicted performance of both the CSTR and the USRB are somewhat unusual. The CSTR exhibits a rather dramatic drop in conversion with increasing dilution rates. This character is not in keeping with that typically seen for CSTR growth on soluble substrates. While there are many feedback mechanisms included in the model, ethanol inhibition of hydrolysis and growth and soluble sugar inhibition of cellulase synthesis and hydrolysis, only the first is a potential explanation for the observed behavior as sugar concentrations in the CSTR are predicted to be extremely low and ethanol inhibition of growth is not unique to insoluble substrates. As ethanol concentration drops with increasing dilution, it is also unlikely that ethanol inhibition of hydrolysis provides a satisfactory explanation. It remains to be established whether the dilution performance exhibited is an intrinsic characteristic of coupled hydrolysis and fermentation or is specific to the organism system modeled.

The one bed model of the USRB is predicted to maintain relatively high conversion (>85%) at residence times down to 3 hours. Such a performance, if realized, would be a significant improvement over other designs proposed for lignocellulose bioconversion. Similar to the CSTR, substrate utilization in the one bed model drops steadily with increasing dilution, albeit at a much slower rate than the CSTR.

The two bed model of the USRB is predicted to respond to increasing dilution rates by maintaining essentially complete utilization until a 5 hour residence time, below which utilization drops rapidly. The two bed model does not predict stable operation at dilution rates as high as the one bed model predicts. It is not absolutely clear at this point whether this behavior results from feedback mechanisms triggered in the two bed model which are not operative in the one bed model or whether this effect is traceable to the stratification of the solids. Further insight into the difference between the one and two bed models may be gained by adding additional solids-concentrating zones, increasing the number numerically modeled from two to five or more.

Figure 5 shows the performance of the one and two bed USRB model conceptions at increasing dilution rates in comparison to the CSTR base case. High substrate utilization is a necessary feature of any economic process for converting lignocellulose to ethanol due to the driving cost of the feedstock. At high utilization, the USRB demonstrates a very significant advantage when compared to the CSTR, with the comparative advantage growing at increasing complete utilization.

Preliminary evaluation of the USRB reactor concept, as demonstrated by these initial modeling efforts, is very promising. When compared to a CSTR, the USRB achieves equal substrate conversion in as little as one tenth the residence time, depending on the conceptual model considered. Further work, both experimental and numerical, is clearly justified by the performance predictions of these models.

Key limitations of the modeling include the lack of experimental determinations of model parameters including cell adherence and reactor voidage. The

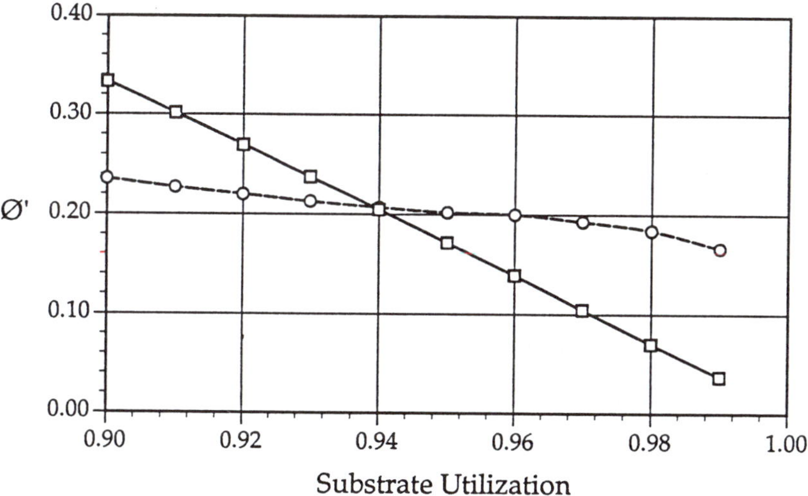

Figure 5. Comparative Dilution for High Substrate Utilization. $\emptyset'$ = D$_{CSTR}$/D$_{USRB}$ for one bed USRB model (O) and two bed USRB model (□).

absence of hydrodynamic considerations and the exclusion of gaseous effects also sharply limit simulation confidence as the fermentation modeled produces essentially equal molar amounts of ethanol and carbon dioxide. Continuing efforts to resolve these issues are ongoing.

BIBLIOGRAPHY

1   Afeyan, N. B. A Mechanistic Study of the *Clostridium thermocellum* Cellulase System. Ph.D. Thesis, MIT. 1987.

2   Ahn, H.-J. Influence of Ethanol and Other Fermentation Products on Growth and Metabolism of *Clostridium thermosaccharolyticum*. M.S. Thesis, Dartmouth College. 1990.

3   Andrews, G. Large-Scale Bioprocessing of Solids. *Biotechnol. Prog.* 6(3):225–230, 1990.

4   Avgerinos, G. C. and D. I. C. Wang. Direct Microbiological Conversion of Cellulosics to Ethanol. *Annual Reports on Fermentation Processes*. 4:165–191, 1980.

5   Bailey, J. E. and D. F. Ollis. "Biochemical Engineering Fundamentals." McGraw-Hill Chemical Engineering Series. Carberry, Fair, Peters, Schowlater and Wei ed. 1986 McGraw-Hill. New York.

6   Bajpai, P., A. Sharma, N. Raghuram and P. K. Bajpai. Rapid production of ethanol in high concentration by immobilized cells of *Saccharomyces cerevisiae* through soya flour supplementation. *Biotechnol. Let.* 10(3): 1988.

7   Bayer, E., R. Kenig and R. Lamed. Adherence of *Clostridium thermocellum* to cellulose. *J. Bacteriol.* **156**(2):818–827, 1983.

8   Bayer, E. A. and R. Lamed. Ultrastructure of the cell surface cellulosome of *Clostridium thermocellum* and its interaction with cellulose. *J. Bacteriol.* **167**(3):828–836, 1987.

9   Beguin, P. Molecular Biology of Cellulose Degradation. *Annual Rev. Microbiol.* **44**:219–248, 1990.

10   Bernardez, T. D. A Kinetic Study of *Clostridium thermocellum* Cellulase: Adsorption and Initial Hydrolysis Rates. M.E. Thesis, Dartmouth College. 1989.

11   Calo, J. M. and G. Hadril. Hydrodynamic Studies of Hydrolysis Reactors. SERI Subcontract Report XK–7–07031–8.

12   Calo, J. M. and G. Hadril. Experimental Studies of the Behavior of Liquid-Fluidized Beds of Aspen Wood Chips. SERI Subcontract Report XK–7–07031–8. 1989.

13   Chambers, R. S., R. A. Herendeen, J. J. Joyce and P. S. Penner. Gasohol: Does it or doesn't it produce positive net energy? *Science.* **206**: 1979.

14   Dale, B. E. *Lignocellulose conversion and the future of fermentation biotechnology.* **5**:287–291, 1987.

15   Energy Information Agency. Monthly Energy Review. DOE/EIA–0035. 1989.

16   Gijzen, Schoenmakers, Caerteling and Vogels. Anaerobic degradation of papermill sludge in a two-phase digestor containing rumen microorganisms and colonized polyurethane foam. **10**(1): 1988.

17   Gilbert, P., D. G. Allison, D. J. Evans, P. S. Handley and M. R. W. Brown. Growth Rate Control of Adherent Bacterial Populations. *Appl. Environ. Microbiol.* **55**(5):1308–1311, 1989.

18   Gonzalez, G., G. Caminal, C. d. Mas and J. Lopez-Santin. An Approach in Mathematical Modeling of an Upflow Packed-Bed Reactor for Enzymatic Hydrolysis of Wheat Straw. *Biotechnol. Bioeng.* **34**:242–251, 1989.

19   Grobicki, A. and D. C. Stuckey. Performance of the Anaerobic Baffled Reactor Under Steady-State and Shock Loading Conditions. *Biotechnol. Bioeng.* **37**:344–355, 1991.

20   Haughney, H. A., D. M. Dziewulski and E. B. Nauman. On material balance models and maintenance coefficients in CSTRs with various extents of biomass recycle. **7**:113–130, 1988.

21   Herendeen, R. and S. Brown. A comparative analysis of net energy from woody biomass. **12**(1):75–84, 1987.

22   Herrero, A. A., R. F. Gomez, B. Snedecor, C. J. Tolman and M. F. Roberts. Growth Inhibition of *Clostridium thermocellum* by Carboxylic Acids: A Mechanism Based on Uncoupling by Weak Acids. *Appl. Microbiol. Biotechnol.* **22**:53–62, 1985.

23   Hogsett, D. A., H.-J. Ahn, P. W. Hill, T.-A. Klapatch, C. R. South and L. R. Lynd. Direct Microbial Conversion: Prospects, Progress, and Obstacles. 13th Symposium on Biotechnology for Fuels and Chemicals. Colorado Springs, CO. 1991.

24 Johnson, E. A., F. Bouchot and A. L. Demain. Regulation of Cellulase Formation in *Clostridium thermocellum*. *J. Gen. Microbiol.* **131**:2303–2308, 1985.

25 Kafarov, V. V., A. Y. Vinarov and L. S. Gordeev. Modeling of Bioreactors. *Intl. Chem. Engr.* **28**(1):14–35, 1988.

26 Keim, C. R. and K. Venkatasubramanian. Economics of Current Biotechnological Methods of Producing Ethanol. TIBTECH. **7**:22–29, 1989.

27 Khan, A. W., S. S. Miller and W. D. Murray. Development of a Two-Phase Combination Fermenter for the Conversion of Cellulose to Methane. *Biotechnol. Bioeng.* **25**:1571–1579, 1983.

28 Klyosov, A. A. Cellulolytic Enzymes. manuscript. 1990.

29 Kobayashi, T., M. P. M. Romaniec, U. Fauth and A. L. Demain. Subcellulosome Preparation with High Cellulase Activity from *Clostridium thermocellum*. *Appl. Environ. Microbiol.* **56**(10):3040–3046, 1990.

30 Kurose, N., J. Yagyu and T. Miyazaki. Evaluation of Ethanol Productivity from Cellulose by *Clostridium thermocellum*. *J. Ferment. Technol.* **64**(5): 1986.

31 L'Italien, Y., J. Thibault and A. LeDuy. Improvement of ethanol fermentation under hyperbaric conditions. *Biotechnol. Bioeng.* **33**:173–182, 1989.

32 Lamed, R. and E. A. Bayer. The Cellulosome of *Clostridium thermocellum*. *Adv. Appl. Microbiol.* **33**:1–46, 1988.

33 Lamed, R., R. Kenig, E. Morag, J. F. Calzada, F. d. Micheo and E. A. Bayer. Efficient Cellulose Solubilization by a Combined Cellulosome-B-Glucosidase System. *Appl. Biochem. Biotechnol.* **27**:173–183, 1991.

34 Lamed, R. J., J. H. Lobos and T. M. Su. Effects of stirring and hydrogen on fermentation products of *Clostridium thermocellum*. *Appl. Environ. Microbiol.* **54**(5):1216–1221, 1988.

35 Lynd, L. R. "Production of ethanol from lignocellulosic materials using thermophilic bacteria: critical evaluation of potential and review." *Advances in Biochemical Engineering and Biotechnology.* Fiechter ed. 1989 Springer-Verlag. Berlin.

36 Lynd, L. R., H.-J. Ahn, G. Anderson, P. Hill, D. S. Kersey and T. Klapatch. Thermophilic Ethanol Production: Investigation of Ethanol Yield and Tolerance in Continuous Culture. *Appl. Biochem. Biotechnol.* 1991.

37 Lynd, L. R., J. H. Cushman, R. J. Nichols and C. E. Wyman. Fuel Ethanol from Cellulosic Biomass. *Science.* **251**:1318–1323, 1991.

38 Lynd, L. R. and H. E. Grethlein. Hydrolysis of Dilute Acid Pretreated Mixed Hardwood and Purified Microcrystalline Cellulose by Cell-Free Broth from *Clostridium thermocellum*. *Biotechnol. Bioeng.* **29**:92–100, 1987.

39 Lynd, L. R., H. E. Grethlein and R. H. Wolkin. Fermentation of Cellulosic Substrates in Batch and Continuous Culture by *Clostridium thermocellum*. *Appl. Environ. Microbiol.* **55**(12):3131–3139, 1989.

40 Manderson, G. J., K. Spencer, A. H. J. Paterson, N. Qureshi and D. E. Janssen. Price Sensitivity of Bioethanol Produced in New Zealand from *Pinus radiata* Wood. *Energy Sources.* **11**:135–150, 1989.

41 Mayer, F., M. P. Coughlan, Y. Mori and L. G. Ljungdahl. Macromolecular

organization of the celluloytic enzyme complex of *Clostridium thermocellum* as revealed by electron microscopy. *Appl. Environ. Microbiol.* **53**(12):2785–2792, 1987.

42   Meyer, C. L. and E. T. Papoutsakis. Continuous and biomass recycle fermentations of *Clostridium acetobutylicum*. *Bioprocess. Engr.* (4):1–10, 1989.

43   Mishra, S., P. Beguin and J.-P. Aubert. Transcription of *Clostridium thermocellum* Endoglucanase Genes *celF* and *CelD*. *J. Bacteriol.* **173**(1):80–85, 1991.

44   Misty, F. R. and C. L. Cooney. Production of Ethanol by *Clostridium thermosaccharolyticum*: I. Effect of Cell Recycle and Environmental Parameters. *Biotechnol. Bioeng.* **34**:1298–1304, 1989.

45   Morag, E., I. Halevy, E. A. Bayer and R. Lamed. Isolation and Properties of a Major Cellobiohydrolase from the Cellulosome of *Clostridium thermocellum. J. Bacteriol.* **173**(13):4155–4162, 1991.

46   Ngian, K.-F. and W. R. B. Martin. Bed Expansion Characteristics of Liquid Fluidized Particles with Attached Microbial Growth. *Biotechnol. Bioeng.* **22**:1843–1856, 1980.

47   Piret. Modeling of a Process for the Conversion of Biomass to Ethanol. M.S. Thesis, MIT. 1985.

48   Poncelet, D., H. Naveau and E.-J. Nyns. Transient Response of a Solid-Liquid Model Biological Fluidised Bed to a Step Change in Fluid Superficial Velocity. *J. Chem. Tech. Biotechnol.* **48**:439–452, 1990.

49   Puls, J. and T. M. Wood. The Degradation Pattern of Cellulose by Extracellular Cellulases of Aerobic and Anaerobic Microorganisms. *Bioresource Technol.* **36**:15–19, 1991.

50   Qureshi, N. and I. S. Maddox. Integration of continuous production and recovery of solvents from whey permeate: use of immobilized cells of *Clostridium acetobutylicum* in a fluidized bed reactor coupled with gas stripping. *Bioprocess. Engr.* **6**:63–69, 1991.

51   Rasmussen, M. A., B. A. White and R. B. Hespell. Improved Assay for Quantitating Adherence of Ruminal Bacteria to Cellulose. *Appl. Environ. Microbiol.* **55**(8):2089–2091, 1989.

52   Robson, L. M. and G. H. Chambliss. Cellulases of Bacterial Origin. *Enzy. Microb. Technol.* **11**:626–644, 1989.

53   Sperling, D. Alternative transportation fuels: An environmental and energy solution. 1989.

54   Srivastava, R. D., I. M. Campell and B. D. Blaustein. Coal Bioprocessing: A Research Needs Assessment. *Chem. Eng. Prog.* (12):45–53, 1989.

55   Srivastava, V. J., K. F. Fannin, D. P. Chynoweth and J. R. Frank. Improved efficiency and stable digestion of biomass in nonmixed upflow solid reactors. *Humana Press.* 1988.

56   Srivastava, V. J. and H. R. Isaacson. Biogasification of community-derived biomass and solid wastes in a pilot-scale solcon reactor. *10th Symposium on Biotech for Fuels and Chemicals.* 1988.

57   Tan, L. U. L., E. K. C. Yu, N. Campbell and J. N. Saddler. Column Cellulose Hydrolysis Reactor: An Efficient Cellulose Hydrolysis Reactor

with Continuous Cellulase Recycling. *Appl. Microbiol. Biotechnol.* **25**:250–255, 1986.

58 Tan, L. U. L., E. K. C. Yu, P. Mayers and J. N. Saddler. Column Cellulose Hydrolysis Reactor: Cellulase Adsorption Profile. *Appl. Microbiol. Biotechnol.* **25**:256–261, 1986.

59 Taya, M., T. Yagi and T. Kobayahi. Anaerobic Cellulose Digestion and Characterization of the Cultivation Process in Tower-Type Fermentor. **63**(4):363–370, 1985.

60 Tsuji, S., K. Shimizu and M. Matsubara. Performance evaluation of ethanol fermentor systems using a vector-valued objective function. *Biotechnol. Bioeng.* **30**:420–426, 1987.

61 Vallander, L. and K.-E. L. Eriksson. Production of Ethanol from Lignocellulosic Materials: State of the Art. *Adv. Biochem. Eng. Biotechnol.* **42**:63–95, 1991.

62 Wiegel, J. and M. Dykstra. *Clostridium thermocellum*: Adhesion and Sporulation While Adhered to Cellulose and Hemicellulose. *Appl. Microbiol. Biotechnol.* **20**:59–65, 1984.

63 Wohrer, W. A horizontal bioreactor for ethanol production by immobilized cells. *Bioprocess. Eng.* (4):57–61, 1989.

64 Wright, J. D. Ethanol from Lignocellulose: An Overview. Energy Prog. **8**(2):71–78, 1988.

Fabrication of High Temperature Superconducting YBa$_2$Cu$_3$O$_{7-x}$ Films on Metallic Substrates by an in situ Processing

Q. X. JIA

Department of Electrical and Computer Engineering
State University of New York at Buffalo-Amherst
Research Supervisor: Dr W. A. Anderson

ABSTRACT

Superconducting $YBa_2Cu_3O_{7-x}$ (YBCO) thin films were deposited on metallic substrate of Hastelloy C–276 using in situ RF magnetron sputtering. Highly c-axis oriented films with a zero resistance temperature of 81.3K and critical current density of $7.4 \times 10^3 A/cm^2$ at 77K were obtained where $BaTiO_3$ was used as a buffer layer. Scanning electron microscopy (SEM) surface scan showed quite good surface morphology of the films. The clear interfaces among different regions were further confirmed by cross-sectional SEM analysis and Auger electron spectroscopy depth profiling.

INTRODUCTION

The discovery of an oxide superconductor with zero resistance temperature exceeding the liquid nitrogen temperature promised wide application in future technology including energy areas or magnets. One of the most desirable configurations for a useful superconductor is as a wire, tape, or cable product. In order to realize the above forms, it is necessary to develop techniques that can make thin or thick films on flexible metallic substrates. The preference of using thin film forms lies in its much higher critical current density. The more controllable and reproducible processing to deposit thin films also makes thin films widely used compared to thick films or bulk forms. Nevertheless, high temperature superconducting thin films, such as YBa$_2$Cu$_3$O$_{7-x}$ (YBCO), on metallic substrates still show relatively poor superconducting properties compared to the films on single crystal substrates, such as SrTiO$_3$[1] and MgO.[2]

Most of the previous work on depositing YBCO on metallic substrates was done using laser ablation.[3-5] Chemical vapor deposition was also tried to deposit YBCO on metallic substrates.[6] To prevent the interdiffusion between metallic substrate and YBCO, MgO,[3] YSZ,[4] Ag,[5] and SrTiO$_3$[6] have been used as a buffer layer. A double layer of YSZ/Pt was also recently tried.[7-8] In this work, we used RF magnetron sputtering to deposit superconducting YBCO thin films on metallic substrates with BaTiO$_3$ as a buffer layer. The advantage of sputtering is the possibility of uniform large area deposition. The choice of BaTiO$_3$ as a buffer layer lies in its good thermal stability over a wide temperature region, compatible thermal expansion coefficient (13x10^{-6}/°C) compared to YBCO (a-axis, 14x10^{-6}/°C; b-axis, 12x10^{-6}/°C) and Hastelloy C–276 (11x10^{-6}/°C), and an excellent lattice match between BaTiO$_3$ (a = 0.399 nm) and YBCO (a = 0.382 nm, b = 0.389 nm).

EXPERIMENTS

A metallic substrate of Hastelloy C–276 (Ni–Cr–Mo alloys) was used as the substrate in this study. The substrate was polished with a finished powder size of 1 μm. A buffer layer of BaTiO$_3$ was deposited on Hastelloy using reactive RF magnetron sputtering from a composite BaTiO$_3$ target, where oxygen was used as a reactive gas. The oxygen partial pressure during sputtering was controlled in the range of 0.5–1 mTorr. The substrate temperature changed from 350°C to 600°C in order to investigate the buffer layer deposition conditions on the properties of YBCO thin films. The YBCO thin film was in situ deposited using RF magnetron sputtering from a single stoichiometric YBCO target at a substrate temperature of around 640°C. A perpendicular arrangement of the substrate with respect to the target was used to avoid resputtering effects and to obtain stoichiometric YBCO thin films. Figure 1 shows the experimental setup for the sputtering. Detailed descriptions of the deposition procedure with

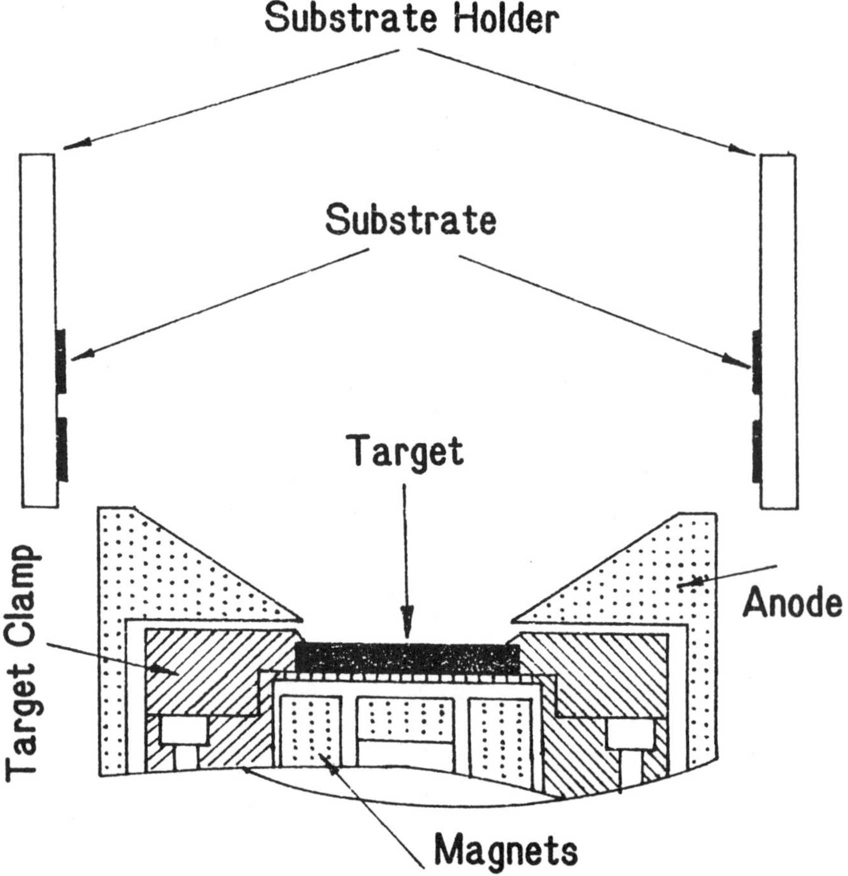

Figure 1.   Sputtering set-up to avoid resputtering effects

this configuration have been published.[9] The YBCO thin films were in situ annealed at 640°C in oxygen ambient (>2 Torr) for 30 min after film deposition. Slow cooling at an even higher oxygen pressure was used to avoid interfacial microcrack formation and also to allow a complete transformation of the YBCO film from tetragonal to orthorhombic.

The structural properties of the films were analyzed by X-ray diffraction (XRD) from a Nicolet/STOE X-ray diffractmeter using Cu Kα radiation. The microstructure of the films was investigated by scanning electron microscopy (SEM) from a Hitachi S–800 scanning electron microprobe. The interface properties among different layers were characterized by Auger electron spectroscopy (AES) from a Perkin-Elmer PHI–660 scanning Auger microprobe. The electrical properties of the films were characterized using standard four-probe resistance vs temperature measurement. The critical current density was determined at a voltage drop across the film of 1 μV/cm.

X20K 1.5μm

Figure 2.   SEM micrograph of the YBCO film on
Hastelloy with BaTiO$_3$ as a buffer layer.

RESULTS AND DISCUSSION

The obvious observation of the YBCO on Hastelloy with BaTiO$_3$ as a buffer
layer was the improvement on the film morphology. The YBCO film surface
appeared shiny black regardless of the BaTiO$_3$ thin film deposition temperature
from 350°C to 600°C although high temperature deposition of BaTiO$_3$ film
resulted in a slightly rough BaTiO$_3$ surface. Figure 2 shows the SEM surface
survey of the YBCO film on Hastelloy with BaTiO$_3$ as a buffer layer. No cracks
or peeled regions were found in a quite large area of the film surface. As shown
in the Figure 3, a sharp interface between different regions can be distin-
guished. This demonstrates less interdiffusion during film growth. Another
very important observation from SEM investigation was the passivation of
grain boundaries on the Hastelloy due to the use of BaTiO$_3$ as a buffer layer.[10]
The groove-like boundaries on the heated Hastelloy could be observed clearly
if YSZ was used as a buffer layer.[7]

X40K 0.75µm

Figure 3. Cross-sectional SEM micrograph of the YBCO on BaTiO3/Hastelloy substrate.

Structural analysis using XRD showed very different texture properties for BaTiO3 and YBCO films, respectively. As shown in Figure 4 (a), the BaTiO3 film deposited at a temperature of 350°C with a thickness of around 150 nm seems to be featureless with an amorphous structure. However, the YBCO film deposited on such a substrate showed highly textured structures as shown in Figure 4 (b). Pure c-axis oriented grain growth perpendicular to the substrate surface can be clearly deduced from Figure 4 (b). The c-axis lattice constant based on XRD data from the (006) diffraction peak was 1.169–1.170 nm.

Surface analysis of the YBCO film using AES surface scan revealed only expected elements as shown in Figure 5. Y, Ba, Cu, and O showed up on the YBCO film surface, where the thickness of the YBCO layer was around 250 nm. This demonstrated BaTiO3 to be an effective barrier layer to prevent the interdiffusion between YBCO and Hastelloy. No serious interdiffusion among different regions was also confirmed from AES depth profiling as shown in

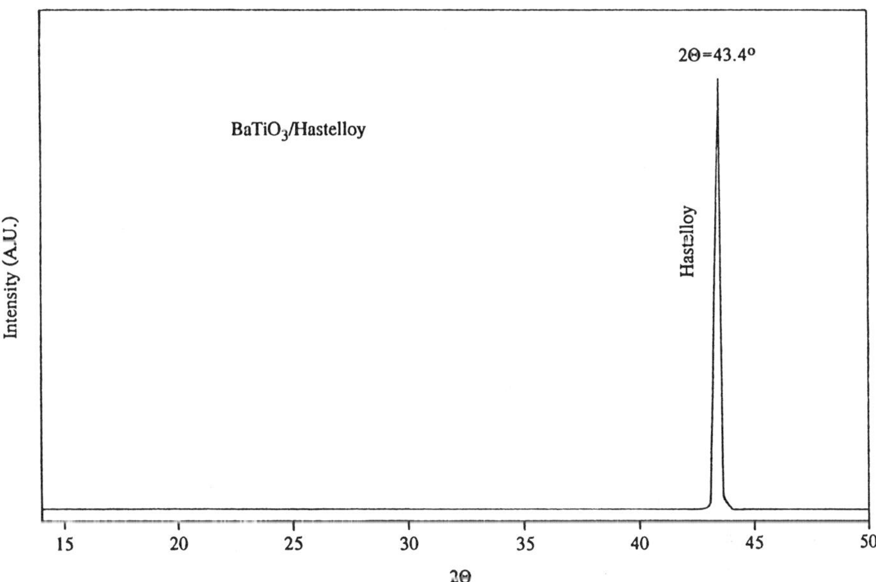

Figure 4. XRD patterns for (a) BaTiO3 on Hastelloy, and (b) YBCO on BaTiO3/Hastelloy.

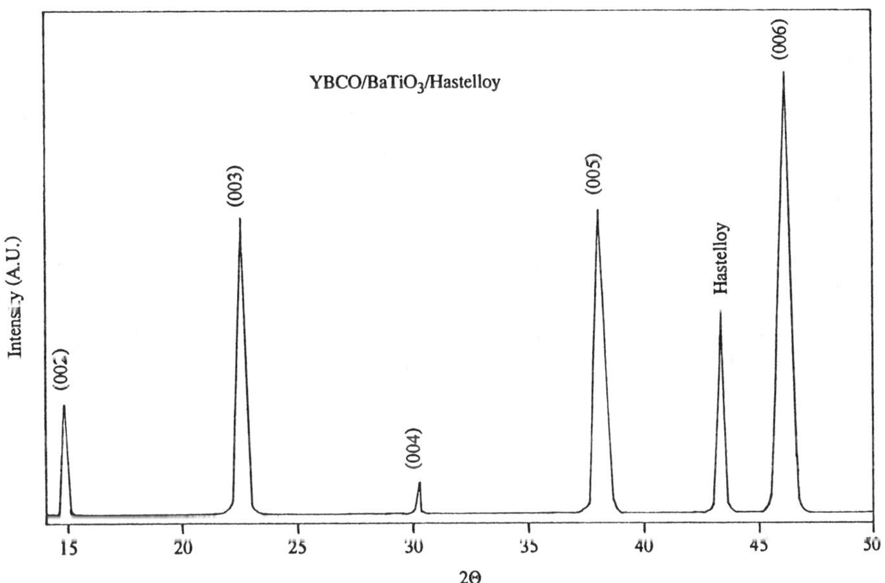

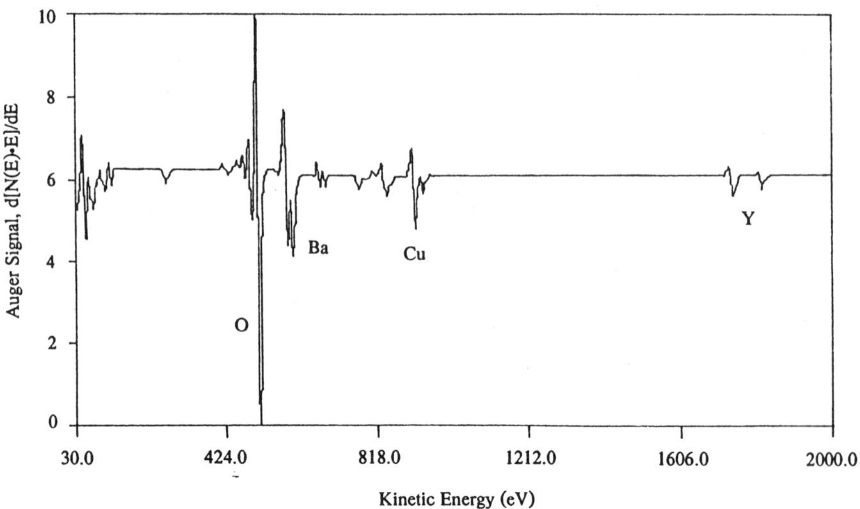

Figure 5. AES surface survey of the YBCO film on BaTiO3/Hastelloy substrate.

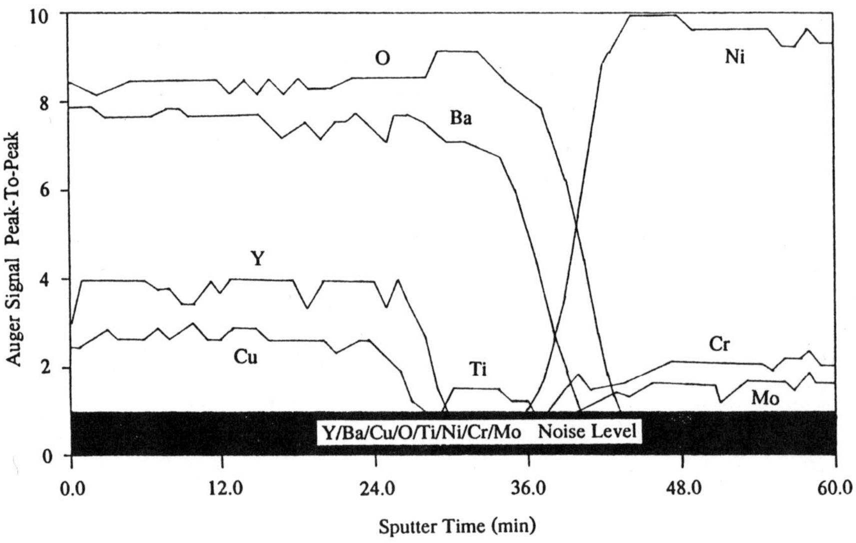

Figure 6. AES depth profiling of the YBCO film on Hastelloy using BaTiO3 as a buffer layer.

Figure 6. The YBCO composition through the film thickness was quite uniform. However, oxidation of the Hastelloy surface might occur during $BaTiO_3$ deposition or if oxygen diffuses through the $BaTiO_3$ layer during YBCO film deposition since a relatively broad transition region appears between $BaTiO_3$ and Hastelloy as demonstrated in Figure 6. The reactive gas of oxygen during either $BaTiO_3$ or YBCO film deposition might react with Hastelloy substrate, in particular, at a relatively high temperature.

Electrical measurement on the YBCO films showed the metallic nature of R–T curves. Films with a thickness of around 250 nm and a buffer layer thickness of 150 nm became superconductive above liquid nitrogen temperature. Our best data for zero resistance temperature of the films was 81.3°K. The critical current density of the film at 77°K was around $7.4 \times 10^3 A/cm^2$. The measurement was done on a film with a width of 3 mm. This result was even better than YBCO on Hastelloy with YSZ as a buffer layer.[4]

SUMMARY

High temperature superconducting YBCO thin films have been deposited on metallic substrate of Hastelloy C–276 using in situ RF magnetron sputtering. The interdiffusion between substrate and YBCO during processing can be effectively reduced using $BaTiO_3$ as a buffer layer. The added advantage of using $BaTiO_3$ as a buffer is the passivation effects on the Hastelloy. Highly c-axis oriented superconducting YBCO thin films with zero resistance temperature of 81.3°K and a critical current density of $7.4 \times 10^3 A/cm^2$ have been obtained in our experiments. Further increasing critical current density and zero resistance temperature of the YBCO films by optimizing the processing parameters is underway in our lab.

ACKNOWLEDGMENT

We wish to thank P. Bush and R. Barone for their technical assistance in AES and SEM analysis. This work was supported by a Link Foundation Energy Fellowship.

REFERENCES

1   X.D. Wu, R.E. Muenchausen, S. Foltyn, R.C. Estler, R.C. Dye, A. Garcia, N.S. Nogar, P. England, R. Ramesh, D.M. Hwang, T.S. Ravi, C.C. Chang, T. Venkatesan, X.X. Xi, Q. Li, and A. Inam, Appl. Phys. Lett. 57, 523(1990)

2   R.H. Moeckly, S.E. Russek, D.K. Lathrop, R.A. Buhrman, J. Li, and J.W. Mayer, Appl. Phys. Lett. 57, 1687(1990)

3   J. Saitoh, M. Fukutomi, Y. Tanaka, T. Asano, H. Maeda, and H. Takahara, Jpn. J. Appl. Phys. 29, L1117(1990)

4   E. Narumi, L.W. Song, F. Yang, S. Patel, Y.H. Kao, and D.T. Shaw, Appl. Phys. Lett. 56, 2684(1990)

5   R.E. Russo, R.P. Reade, J.M. McMillan, and B.L. Olsen, J. Appl. Phys. 68, 1354(1990)

6   T. Yamaguchi, S. Aoki, N. Sadakata, O. Kohno, and H. Osanai, Appl. Phys. Lett. 55, 1581(1989)

7   E. Narumi, L.W. Song, F. Yang, S. Patel, Y.H. Kao, and D.T. Shaw, Appl. Phys. Lett. 58, 1202(1991)

8   J. Saitoh, M. Fukutomi, K. Komori, Y. Tanaka, T. Asano, H. Maeda, and H. Takahara, Jpn. J. Appl. Phys. 30, L898(1991)

9   Q.X. Jia and W.A. Anderson, Appl. Phys. Lett. 57, 304(1990)

10 Q.X. Jia and R. Barone, private communication

Modeling of Quasioptical Josephson Oscillator[†]

BIN LIU

Department of Electrical Engineering
University of Rochester
Research Supervisor: Dr Michael J. Wengler

[†] This paper will be published in *IEEE Trans. on Applied Superconductivity*, Vol. 1, Dec., 1991

ABSTRACT

We have developed a computer model of a Josephson tunnel junction embedded in a general circuit with frequency dependent impedance using the harmonic balance method. This model has been applied to the analysis of a two dimensional Josephson junction array with integrated coupling structures, called a quasioptical Josephson oscillator. The simulations are done for a junction with dipole, slotline and bowtie antennas. The results show that the junction with bowtie antenna gives the best performance and the output power from an array of 4000 junctions can reach 25.7 μW at a frequency as high as 1091 GHz for niobium junctions deposited on a 0.207 mm thick quartz substrate.

1. INTRODUCTION

A Josephson junction is a voltage tunable oscillator which is able to operate up to the gap frequency.[1] This frequency is about 1 THz for lead-alloy and 1.5 THz for niobium. This makes a Josephson oscillator made out of lead-alloy or niobium a natural candidate for a millimeter or submillimeter wave generator. Such an oscillator will find many applications, e.g., as a local oscillator for SIS receiver systems.[2]

There are four problems commonly associated with a Josephson oscillator. The first is very low output power from a single junction. For a junction with a dc bias voltage V, the maximum output power is $I_C V/2$. For I_C in the range of 10 to 1000 μA and bias voltage V at 1 mV, this power is 5 to 500 nW. The second is poor coupling between a Josephson oscillator and the outside world. All Josephson oscillators to date employ resistively shunted junctions with small shunt resistances typically on the order of 1 Ω.[3-6] If a SIS tunnel junction, which has a normal state resistance typically in the range from 50 to 100 Ω, is used as a load, there will be a big mismatch between a Josephson oscillator and the load. The third is the broad radiation linewidth of a single junction.[3] The fourth is the harmonic content in the oscillation from a Josephson oscillator.

Various attempts have been made in overcoming these difficulties to build a practical Josephson oscillator.[3-6] To obtain usable output power, power must be combined from many junctions. A natural way is to put the junctions in an array and let them oscillate in phase. One dimensional Josephson junction arrays have been extensively studied over the past decade.[3-5] The state of the art is an one dimensional array of 40 junctions connected with a stripline. The measured output power is about 1 μW in the 350 GHz to 450 GHz range.[3,4] Putting junctions in a serial array also improves the impedance match between oscillators and loads, and reduces the radiation linewidth.

To reduce the harmonic content, the common practice to date is to bias Josephson junctions more toward the linear region, typically above $I_C R$. For bias voltages from $I_C R$ to $2I_C R$, a resistive shunted junction still diverts 17% to 5.4% of the total rf power into higher harmonics.

Recently we proposed a new scheme of power combining for Josephson oscillators[7] It is a two-dimensional junction array with integrated coupling structures called the Quasioptical Josephson Oscillator (QJO). In our design of the QJO, we took a different approach to the problems mentioned above. We use a two-dimensional array instead of an one-dimensional one and bias all junctions in parallel to ensure all junctions oscillate with the same frequency. We use unshunted SIS tunnel junctions instead of resistively shunted junctions. In this way we reduce the rf power lost to junction resistors, and make the array close to a match to a load. The most unique feature in our design of the QJO is the tuning circuit naturally incorporated in it. It is a resonant cavity behind the array. When the Josephson oscillation hits one of the resonant modes of the cavity, resonance occurs only for the first harmonic oscillation but not for higher harmonics. Harmonic content is therefore greatly reduced in the output compared to that from resistively shunted junctions. In this paper we present an analysis of this scheme of power combining using Josephson tunnel junction oscillators.

Fig. 1 shows a functional schematic of the QJO. The junction array is fabricated on one side of a substrate. The other side of the substrate is metalized to form a cavity behind the array. The cavity functions in two ways. In one way it tunes out the junction capacitance. In the other way it provides a locking mechanism to have all junctions oscillate in phase.

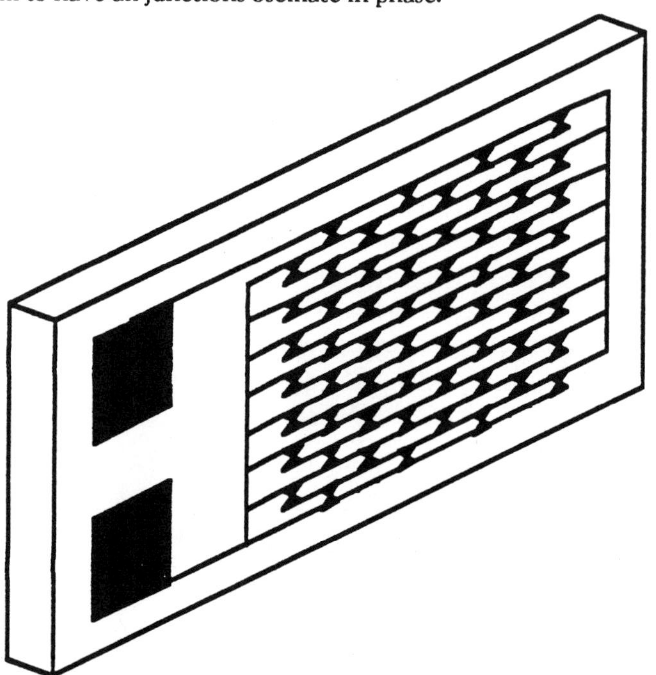

Fig. 1. A schematic design of the quasi optical Josephson oscillator with integrated bowtie antennas. There is one junction at the center of each bow-tie dipole antenna. The dc wiring is shown so that all junctions are biased in parallel.

2. COMPUTER MODELING

2.1 SIMULATION ALGORITHM

The Josephson effect in a superconducting tunnel junction is described by Josephson relations,[8]

$$I = I_c \sin\phi(t) \tag{1}$$

$$\frac{d\phi(t)}{dt} = \frac{2e}{\hbar} V(t) \tag{2}$$

The current response function is:

$$I(t) = I_C \sin\left(\frac{2e}{\hbar} \int^t V(\tau)d\tau\right).$$

The behavior of a Josephson junction is commonly modeled with the RSJ model.[9,10] In this model the analytical solution can be obtained only for zero capacitance junctions. Numerical methods are used for junctions with finite capacitance.

There has been a large amount of work done on simulation of Josephson junctions in the time domain.[11-13] These are difficult to apply when a junction is embedded in a circuit with complicated frequency dependent impedance such as the antennas we propose. For simulation in frequency domain, we adopted the harmonic balance method, which has been used in the analysis of microwave nonlinear circuits.[14]

In our simulation the following normalized variables are used:

resistance:	$r = R/R_N$
current:	$i = I/I_C$
voltage:	$v = V/V_N,\ V_N = I_C R_N$
capacitance:	$c = C/C_N,\ C_N = h/2eI_C R_N^2$
inductance:	$l = L/L_N,\ L_N = h/2eI_C$
time:	$t = T/T_N,\ T_N = h/2eI_C R_N$
frequency:	$f = F/F_N,\ F_N = 2eV_O/h$
power:	$p = P/P_N,\ P_N = I_C^2 R_N$

Here I_C is the critical current of the junction, and R_N is the normal state resistance of the tunnel junction. In normalized units, $f_J = v_O$ and the Josephson relations become:

$$i(t) = \sin(\phi(t)) \tag{3}$$

$$\frac{d\phi(t)}{dt} = 2\pi v(t) \tag{4}$$

The algorithm for simulation of a Josephson junction with embedding impedance $Z(\omega)$ as shown in Fig. 2 is as follows:

(1) Assume the junction voltage $v_J(t)$. We always set $v_J(t) = v_O$ as the initial value, where v_O is the dc bias voltage.

(2) Fourier transform $v_J(t)$ to $v_J(\omega)$.

(3) Use the Fourier transform of eqn. (4)

$$-j\omega\phi(\omega) = 2\pi v(\omega)$$

to find $\phi(\omega)$.

(4) Inverse Fourier transform $\phi(\omega)$ to $\phi(t)$.
(5) Use eqn. (3) to find the junction current in time domain $i_J(t)$.
(6) Fourier transform $i_J(t)$ to $i_J(\omega)$.
(7) Find the voltage across the linear embedding circuit
$$v_L(\omega) = i_L(\omega)z(\omega) = (-i_J(\omega))z(\omega).$$
(8) Compare $v_L(\omega)$ with $v_J(\omega)$. If the magnitude of the error is larger than the preset precision parameter, go back to step (3) and use $v_L(\omega)$ as $v_J(\omega)$. Otherwise the simulation is done.

In all the iterations, the dc component of the junction voltage is kept constant at v_O, the dc bias voltage. We do 128 point Fourier transforms. The fundamental oscillation frequency corresponds to the fourth point in this array, so 32 harmonics of the fundamental are included in our calculations.

Tested for a resistively-shunted junction (RSJ) circuit, the above procedure converges only for $V_O > 0.2\ I_C R_N$. Convergence below this voltage could not be achieved even with inclusion of (1) more harmonics or (2) a damping factor in the iteration, or (3) double precision arithmetic.

We were able to improve convergence at low bias voltages by including the identity network of Hicks and Khan,[14] shown in Fig. 2. The positive admittance is added to the linear side of the calculation where convergence is not an issue. The corresponding negative impedance serves to stabilize the non-linear or Josephson junction side of the iterative calculation, where convergence is a problem. Though the convergence rate is slower by adding in the identity network, convergence can be achieved for almost all bias voltages for a resistively shunted junction.

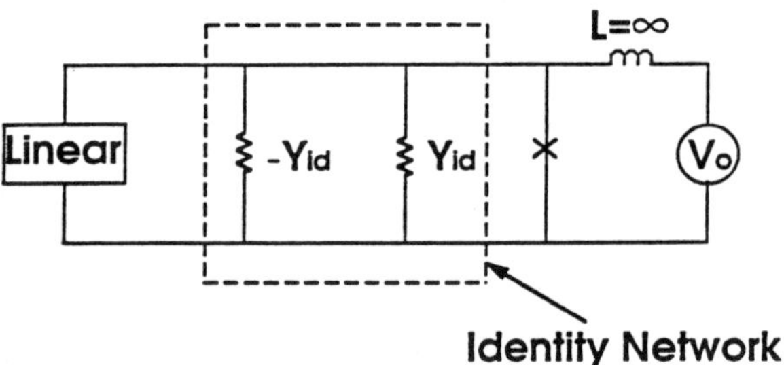

Fig. 2. The junction circuit model for our simulation program. The identity network is introduced to achieve wide range convergence.

Finally, we compare the results of our simulation program with the analytic solution for the dc current-voltage relation of an RSJ circuit. The outcome of the simulation program is within 0.05% of the analytical solution.

2.2 CIRCUIT MODEL OF THE QJO

2.2.1 A tunnel junction

A real tunnel junction in the presence of rf radiation can be modeled as an ideal Josephson junction shunted by a parasitic capacitance and resistance as shown in Fig 4. The capacitance arises from the superconductor-insulator-superconductor sandwich structure of a tunnel junction. A typical value for a micron size junction is about 50 fF. The resistance arises from the quasi-particle tunneling across the junction in the presence of rf radiation. This quasi-particle tunneling current is close to zero at small bias voltages and has a finite value at bias voltages above a certain value, where the first photon step due to rf radiation just overlaps with the Josephson oscillation. This voltage is 1/3 V_{GAP}. In general the parasitic resistance is a voltage dependent variable.[15] As an approximation, however, we assume the resistance is a constant for all the bias voltages and equals the junction normal state resistance R_N.

Our QJO's will be fabricated using both lead alloy and niobium technology on quartz and silicon substrate. For a typical niobium junction, the junction area A = 1 μm^2, I_C = 40 μA, R_N = 75 Ω, C = 50 fF. The normalization units for this junction are: V_N = 3 mV, F_N = 1452 GHz, C_N = 9.2 fF, L_N = 52 pH in our simulation.

2.2.2 Antenna arrays

(a) **Dipole antenna array:** The simplest coupling structure is the square grid as shown in Fig. 3. This structure has been successfully used by Rutledge's group at Caltech for phase shifters,[16] multipliers,[17] and millimeter wave oscillators[18,19] using Schottky diodes and HEMT's. The junctions are located on the center of the vertical strips. These strips are dipole antennas and form an antenna array. We refer to the junction-containing strips as "leads." The horizontal strips are necessary to provide the bias to the junctions.

To simplify the analysis we assume the array is infinite in extent. We also assume all the junctions in the array oscillate in equal phase and amplitude. The phase locking problem of the QJO is not investigated in this paper.

Due to symmetry, the electrical field is perpendicular to the dc bias line and in the planes parallel with the array. We can put electrical walls (perfect conductors) and magnetic walls around unit cells to form a wave guide as shown in fig. 3 without changing anything. Since all unit cells are identical, the analysis of the junction array now is reduced to that of a single unit cell. A unit cell can then be modeled as an equivalent transmission-line circuit as shown in Fig. 4(a). The free space and the dielectric space on both sides of the junction array are modeled as transmission lines with characteristic admittance Y_O and Y_S respectively, where Y_O = 1/377 Ω and Y_S = n/377 Ω with n the index of refraction of the substrate material. The mirror on the back of the substrate is a short to the dielectric space transmission line. The lead is represented by an inductance in series with the Josephson junction. The lead inductance can be calculated using the quasi-static method,[20] which gives:

$$L=\frac{\mu_o a}{2\pi}\ln\left\{\csc\left[\frac{\pi w}{2a}\right]\right\},\tag{5}$$

where a is the length and w is the width of the lead. This equivalent circuit model has been confirmed experimentally.[21,22]

In our array design of the square grid structure, the spacing between two adjacent junctions is 60 μm and the width of the inductive lead is 4 μm. Using the quasi-static formula (5) the lead inductance is calculated to be 27 pH.

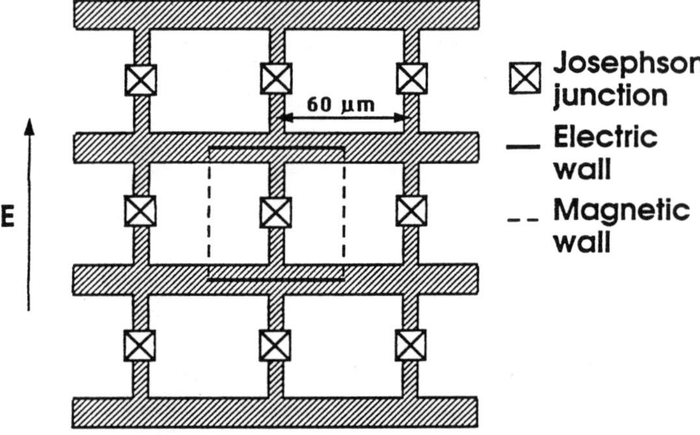

Fig. 3. Square grid. Josephson junctions are at the center of vertical strips. The horizontal strips are bias lines. The electrical field is perpendicular to the bias lines. The unit cell is defined by boundary conditions on the symmetry lines. The solid lines on top and bottom represent electrical walls, and the dashed lines on the sides represent magnetic walls.

(b) Slotline antenna array: The circuit model for a unit cell of a slotline antenna array is same as that for a unit cell of a dipole antenna array but with smaller lead inductance. In our design of the array with slotline antenna, the lead length is reduced to 1/6 of the periodic spacing between junctions. The inductance is then 3.8 pH calculated using the quasi-static formula. With reduced distance between the up and bottom electric pads, the capacitance between them needs to be taken into consideration. Using the quasi-static formula:[20]

$$C=\frac{2a}{\pi}\,\varepsilon_o\left(\frac{1+\varepsilon_r}{2}\right)\ln\left\{\csc\left[\frac{\pi g}{2a}\right]\right\},\tag{6}$$

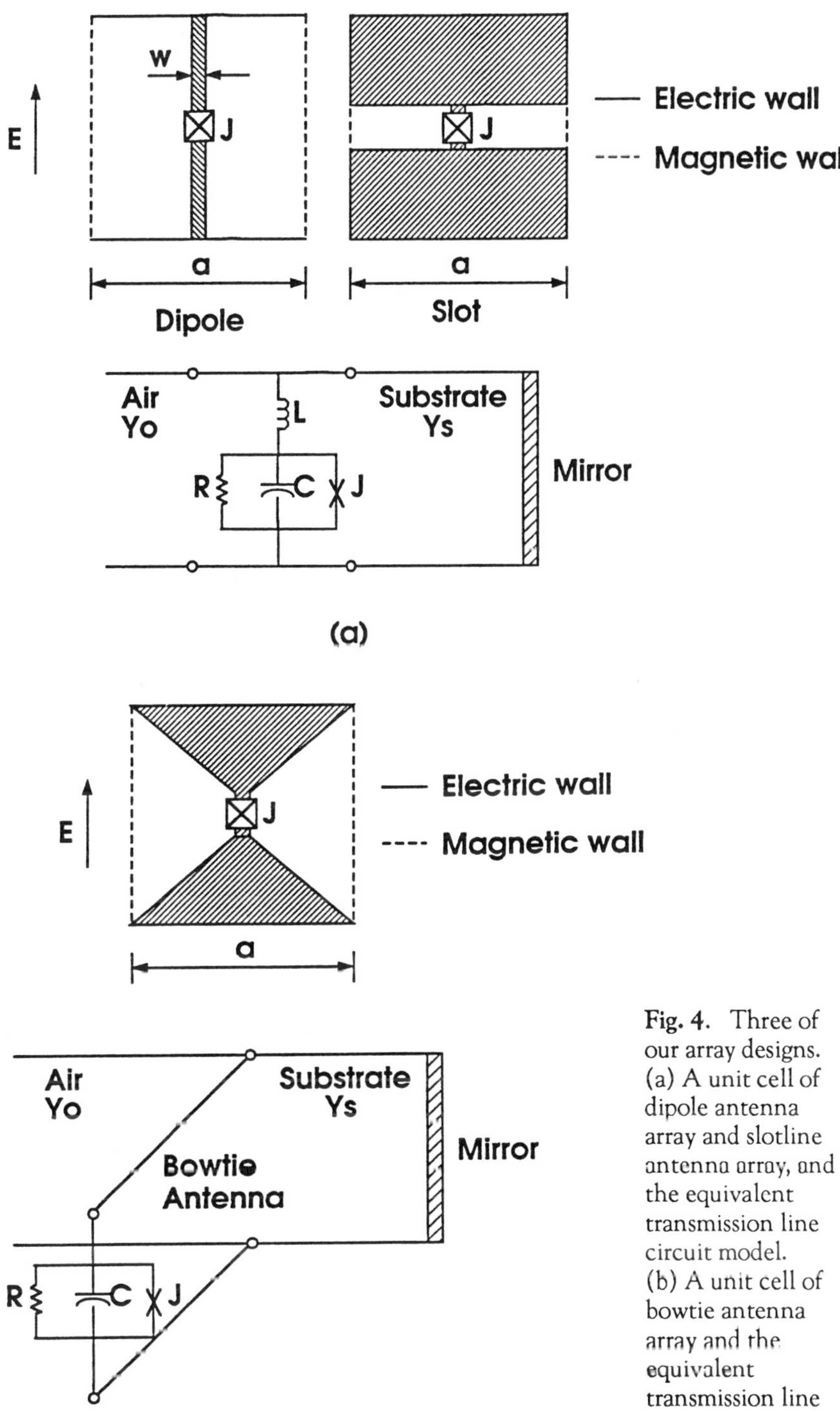

Fig. 4. Three of our array designs. (a) A unit cell of dipole antenna array and slotline antenna array, and the equivalent transmission line circuit model. (b) A unit cell of bowtie antenna array and the equivalent transmission line circuit model.

where ε_r is the dielectric constant of the substrate, a is the length and g is the width of the gap between two metal plates, the capacitance is calculated to be 1.1 pF. This capacitance is in parallel with the junction capacitance. Since this capacitance is small compared to the junction capacitance, it is neglected in our simulation.

(c) **Bowtie antenna array:** A single bowtie antenna is a very broadband antenna. Its driving point impedance is purely real.[23] An infinite bowtie antenna can be thought as an infinite transmission line with characteristic impedance determined by the bow angle. In the circuit model for a unit cell of a bowtie antenna array, the bowtie antenna is represented by a transmission line with certain characteristic impedance and electrical length (see Fig. 4(b)). In the microwave modeling of bowtie antenna arrays done by Weikle et al., he confirmed this model for a array of regular bowtie antennas with 8 mm periodic spacing for frequencies from 2 to 18 GHz.[21] The characteristic impedance and the electrical length of the transmission line he obtained are 78 Ω and 23° at 10 GHz respectively. Scaling down to the actual dimension in our design of the QJO, the corresponding frequency is 1333 GHz.

(d) **Ideal coupling:** The intention of using slotline and bowtie antennas is to reduce the lead inductance and improve the coupling efficiency between the junction and the transmission lines. The limit is ideal coupling, in which the Josephson junction is coupled to the transmission lines directly. This situation is of interest to us because we then know how much more we can further improve our designs.

(e) **Ideal coupling with reduced height:** Even at ideal coupling the power is not completely coupled out to free space because of the mismatch between the junction resistance and the free space impedance. The typical value of a tunnel junction resistance is from 50 to 100 Ω. There are two ways to make them match. One way is to raise the junction parasitic resistance. But higher junction resistance means smaller critical current and thus less power from a junction. Another way is to reduce the free space impedance a junction sees. This can be done by reducing the height of the unit cell along electrical field. The idea is similar to that of the reduced height wave guide. The impedance is then 377r Ω, where r is the height to width ratio.

2.3 SIMULATION RESULTS

Simulations are done for the junction with all the coupling structures discussed above. The results are shown in Figs. 5, 6, 7, and summarized in Table 1. The substrate thickness is chosen so that the shorted transmission line is one-half wavelength long at 363 GHz. This thickness is 0.207 mm for a quartz substrate. The resonances of the circuit occur at frequencies which are slightly offset from 363 GHz and its harmonics. The offset from the exact frequencies occurs because the shorted transmission line must tune out the junction's capacitive susceptance at resonance.

At these resonances the load to the Josephson junction is resistive and the photon emission process of Cooper pair tunneling is enhanced. Away from these resonances the reactive part of the load dominates the load to the junction and suppresses the photon emission process. The current plotted in part (a) of Figs. 5, 6, 7 is the dc current flowing through the junction. The total rf power the junction radiates is simply this dc current times the dc bias voltage, since the Josephson (pair) current in the junction is nondissipative.

We can also directly calculate the part of the rf power that is at the fundamental radiation frequency. This is shown in Table 1. More than 98.6% of the rf power from the junction goes into the fundamental at the resonances for all coupling structures investigated. This is because the shorted transmission line tunes out the junction capacitance at the fundamental, but it presents a large susceptance to the junction at higher harmonics.

(a) Dipole antenna array: Fig. 5(b) shows the rf power a junction in the QJO with a dipole antenna array radiates into free space at the first harmonic and Fig. 5(c) is a blowup of the resonant peak around 0.75 mV. The peak power is 1.48 nW at 0.6906 mV. For an array of 4000 junctions, the output power is 5.92 μW at 334 GHz. The 3dB linewidth of the peak is 3.2%.

The simulation result shows that the output power of the QJO with a dipole antenna array rolls off rapidly toward high frequencies. This is because of the lead inductance between the junction and the transmission lines. This lead inductance is not a problem in the operation of the devices build by Rutledge's group at Caltech using the same coupling structure, because they operate the devices at low frequencies. At high frequencies, however, the impedance of the lead inductance is more important in the circuit and prevents the effective coupling between the active device and the transmission lines. In the QJO with the grid coupling structure, the lead inductance prevents the shorted transmission line from tuning out the junction capacitance. Since we intend to operate the QJO at high frequencies (above 500 GHz), the lead inductance needs to be reduced or eliminated. A straight forward way of reducing the lead inductive reactance is to reduce the length of the lead, since the inductance is almost proportional to the lead length. This leads us to the slotline antenna as shown in Fig. 4(a).

(b) Slotline antenna array: The simulated result for the QJO with a slotline antenna array is shown in Fig. 6. We can see that the performance of the junction at high frequency is improved. At frequency of 716 GHz, the rf power coupled out to free space at the first harmonic from a junction is 2.40 nW. For an array of 4000 junctions, the output power is 9.60 μW. At higher frequencies, however, the output power drops because even the small lead inductance turns

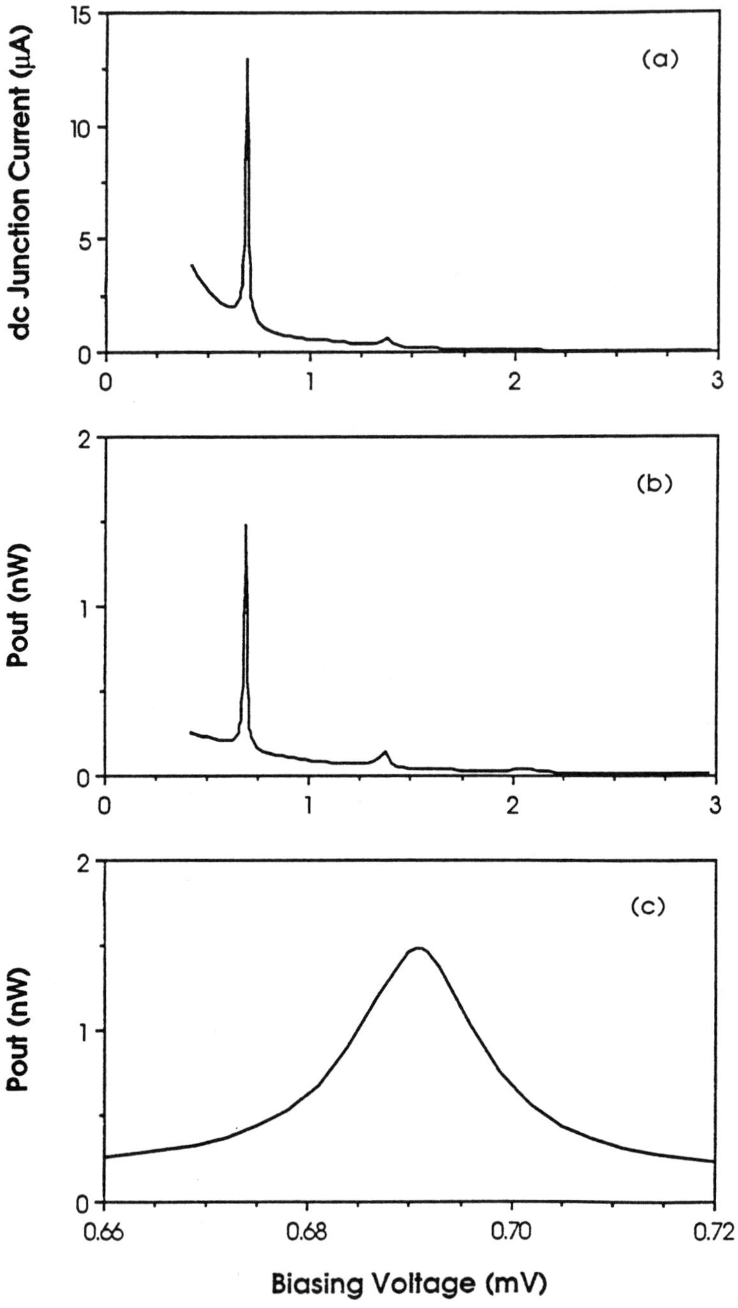

Fig. 5. Simulated results for a unit cell of the junction array with dipole antennas. (a) dc current flowing through the junction vs. bias voltage. (b) The power at the first harmonic the junction radiates vs. bias voltage. (c) Blowup of the first resonance peak in the graph of (b).

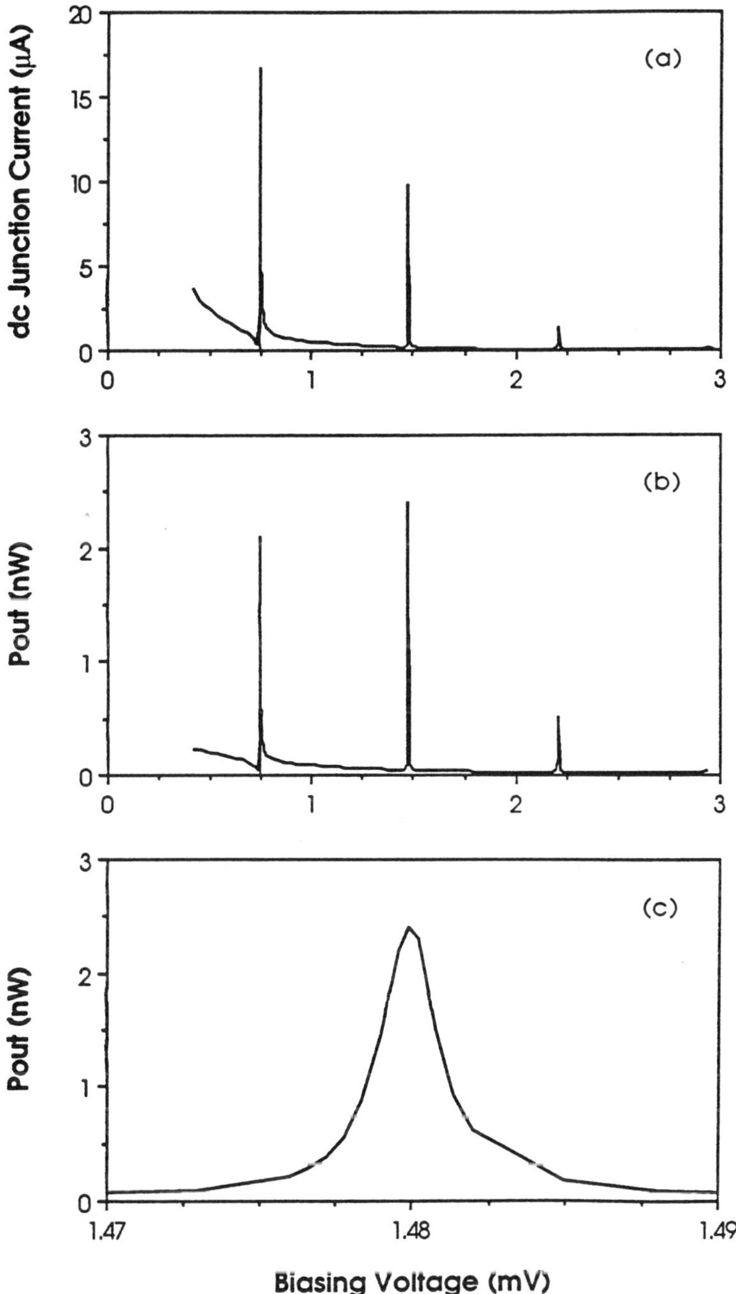

Fig. 6. Simulated results for a unit cell of the junction array with slotline antennas. (a) dc current flowing through the junction vs. bias voltage.
(b) The power at the first harmonic the junction radiates vs. bias voltage.
(c) Blowup of the second resonance peak in the graph of (b).

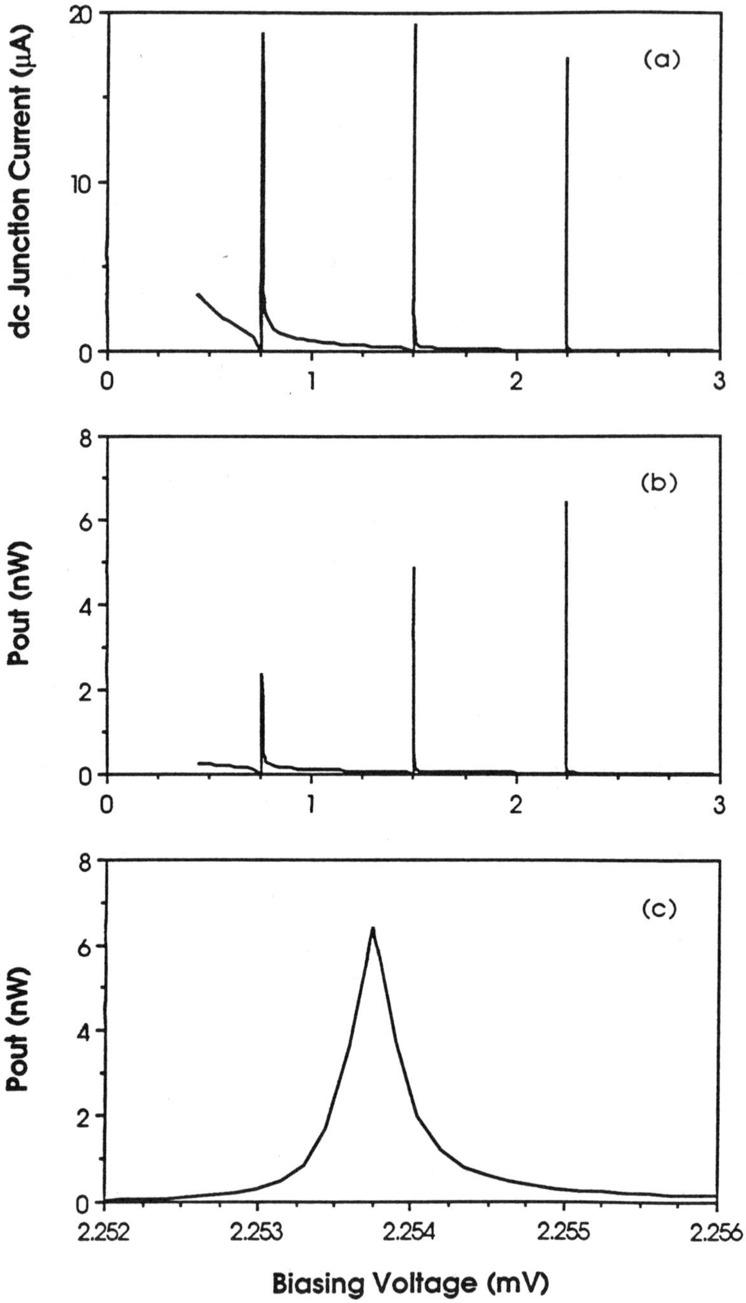

Fig. 7. Simulated results for a unit cell of the junction array with bowtie antennas. (a) dc current flowing through the junction vs. bias voltage. (b) The power at the first harmonic the junction radiates vs. bias voltage. (c) Blowup of the third resonance peak in the graph of (b).

out to have big enough impedance to block the effective coupling between a junction and the transmission lines. At frequency of 1072 GHz, the rf power from a single junction is only 0.518 nW. For an array of 4000 junctions, the total output power is 2.07 μW.

(c) **Bowtie antenna array:**   Another way of reducing the lead inductance is to flare the lead out into a triangle and form a bowtie antenna. The simulation result for the QJO with a bowtie antenna array is shown in Fig. 7. The performance of a junction at high frequencies is greatly improved. At the frequency of 729 GHz, the power coupled out from one junction is 4.85 nW. For an array of 4000 junctions, the total output power is 19.4 μW. At the higher frequency of 1091 GHz, the bowtie antenna coupling is close to the ideal coupling between the junction and transmission lines (see Table 1). The output power from a 4000 junction array is 25.7 μW. A comparison between the output from junctions with dipole, slotline and bowtie antennas are listed in Table 1. Among three coupling structures, the bowtie antenna gives the best performance over the entire frequency range and especially in high frequency region, where the output from a junction with dipole antenna is diminishing and the output from a junction with slotline antenna is small.

(d) **Ideal coupling:**   The simulation results for the ideal coupling between a junction and the transmission lines are shown in Fig. 8. The solid curve is the

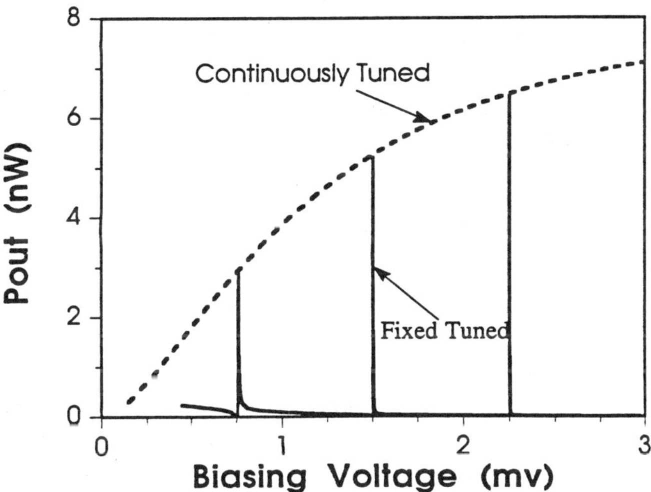

Fig. 8. Simulated results for the ideal coupling between a junction and the transmission lines. The solid curve is the output power of the junction with fixed-tuned cavity. The dashed curve is the output power of the junction with a continuously-tuned cavity tuned for the optimum power output.

output power of the junction with fixed-tuned cavity. The dashed curve is the output power of the junction with a continuously-tuned cavity tuned for the optimum power output. We expect the output power from a junction ideally coupled to the cavity and to free space should increase all the way up to the gap frequency.

(e) Ideal coupling with reduced height: The simulation results for ideal coupling between junctions and transmission lines with reduced height structure are listed in Table 1. A height to width ratio of 1/5 is used. The free space impedance to a reduced height unit cell is 75.4 Ω. The power coupled out to free space from a junction is more than doubled than that from a junction of ideal coupling with regular coupling structure at all resonances.

	PEAK 1				PEAK 2				PEAK 3			
	V (mV)	ΔV ($\times 10^{-3}$)	P (nW)	ΔP (x%)	V (mV)	ΔV ($\times 10^{-3}$)	P (nW)	ΔP (x%)	V (mV)	ΔV ($\times 10^{-3}$)	P (nW)	ΔP (x%)
DIPOLE	0.6906	32.1	1.48	98.9	1.380	58.0	0.139	98.6	NO PROMINENT PEAK			
SLOTLINE	0.74769	4.3	2.09	100	1.4799	1.49	2.40	99.5	2.2158	2.26	0.518	99.5
BOWTIE	0.76125	5.1	2.37	99.8	1.5057	2.2	4.85	100	2.2539	0.133	6.42	99.1
IDEAL	0.76107	3.3	2.90	98.0	1.50561	0.50	5.21	98.8	2.25375	0.17	6.42	99.1
IDEAL(r.h.)	0.8022	26.9	8.67	98.6	1.5276	3.93	12.1	99.1	2.2686	1.32	13.4	99.3

Table 1. A comparison between the simulated results of Josephson junctions with the five different coupling structures described in section 2.2.2. V is the junction dc bias voltage. ΔV is the 3dB width of the peaks in the power-voltage curves. P is the power coupled out from a single junction in the array to free space. ΔP is the ratio of the rf power at the fundamental to the total rf power emitted by the oscillator.

2.4 DISCUSSION

(a) **Coupling efficiency and the output power:** A Josephson tunnel junction is not an effective oscillator because the parasitic capacitance shunts the oscillation. By coupling the junction to a resonant cavity the capacitance can be tuned out. This tuning could be impeded if the coupling is not efficient. In the operation of the QJO this impedance comes from the lead inductance. It is our main target in designing different coupling structures to reduce this inductance. From Table 1 we can see that the output power from the QJO improves with increasing coupling efficiency between junctions and the resonant cavity and free space. Our simulation results clearly show that the bowtie antenna array is the favorable choice among the three coupling structures discussed above, especially for high frequency operation.

(b) Radiation linewidth: A Josephson oscillator is notorious for its broad radiation linewidth. For a single tunnel junction, the 3dB radiation linewidth is

$$\Delta v = \pi \left(\frac{R_D}{\Phi_O}\right)^2 S_I(0), \tag{7}$$

where R_D is the dynamic resistance of the junction and

$$S_I(0) = 2e\left[I_{QP}(V)\coth\frac{2eV}{k_BT} + 2I_P(V)\coth\frac{eV}{k_BT}\right]$$

is the current noise spectral density at a bias voltage V and operation temperature T.[24] The equation (7) has two familiar limiting forms: Johnson noise form at low bias voltages and shot noise form at high bias voltages. Since $k_BT/e = 0.36$ mV at $T = 4.2$ K, for bias voltages greater than about 1 mV, (7) becomes:

$$\Delta v = \pi \left(\frac{R_D}{\Phi_O}\right)^2 2e[I_{QP}(V) + 2I_P(V)].$$

For N phase-locked junctions biased in parallel, the dynamic resistance scales as R_D/N, while the current noise scales as $NS_I(0)$. The 3dB radiation linewidth of a QJO without the resonator cavity is then

$$\Delta v = \frac{\pi}{N}\left(\frac{R_D}{\Phi_O}\right)^2 2e[I_{QP}(V) + 2I_P(V)].$$

For an array of 4000 junctions with $R_D = 75\ \Omega$ operated at 4.2 K, Δn is 6.6 MHz. We take $R_D = R_N$ here. In fact, a QJO biased just below the resonanat peaks shown in figs. 5, 6 and 7 will have R_D much less than R_N, and therefore a narrower linewidth than 6.6 MHz. This is about two orders of magnitude smaller than the smallest width of the resonant peaks of the power-voltage curves of the QJO with different coupling structures. Even though the width of a resonant peak can be reduced by operating at a higher resonant mode of the cavity, such a measure may not be needed for an array of a large number of junctions since the parallel biasing scheme has cut the linewidth dramatically.

(c) Upper frequency limit of the QJO: There are two factors which determine the upper limit of operation frequency of the QJO. One is the rapid roll off of the Josephson critical current above the gap frequency.[25] This issue is related to the Riedel singularity and we plan to investigate this problem further in experiment. Another is the resonance of the antenna array. Like a single antenna element, the impedance of an antenna in an array hits a resonance when $\lambda_m/2$, where λ_m is the wavelength calculated from the mean dielectric constant, is close to the dimension of the antenna, and the circuit models we use above are no longer valid[21] For our array design, the frequency of the first resonance is about 1581 GHz, which is above the frequency range our simulations are done.

3. CONCLUSION

From the simulation for various coupling structures we showed: (1) The square grid used for the coupling structure in the microwave range is adoptable for low frequency operation of the QJO. (2) The lead inductive reactance is detrimental to the high frequency operation of the QJO and needs to be minimized. (3) a bowtie antenna array is the best choice for high frequency operation. (4) For reducing the internal loss to the junction resistors and raising the output power of the QJO, the reduced height structure should be used. (5) The radiation linewidth of the QJO is greatly reduced by using parallel biasing scheme. The linewidth scales with $1/N$ with N being the number of junctions. (6) The harmonic content in the radiation from the QJO is very low. In all cases more than 98.6% of the total power is in the fundamental oscillator frequency.

The 4,000 junction array simulated in this paper can produce 25.7 μW at 1091 GHz in the bowtie array. This assumes that the junctions are all mutually phase-locked to the substrate radiation mode, a condition we do not investigate in this paper. Arrays like those simulated here are currently being fabricated. A comparison between experimental and simulated results will be presented later.

ACKNOWLEDGEMENT

This work is supported by grant AFOSR–90–0233 from the Air Force Office of Scientific Research. One of us (B. Liu) gratefully acknowledges the support of a Link Energy Foundation Fellowship.

REFERENCES

1   R.P. Robertazzi and R.A. Buhrman, "Josephson terahertz local oscillator," *IEEE Trans. Magn.*, vol. 25, pp. 1384–1387, 1989.

2   M.J. Wengler, D.P. Woody, R.E. Miller and T.G. Phillips, "A low noise receiver for millimeter and submillimeter wavelengths," *Intl. J. of IR and MM waves*, vol. 6, pp. 697–706, 1985.

3   A.K. Jain, K.K. Likharev, J.E. Lukens, and J.E. Sauvageau, "Mutual phase-locking in Josephson junction arrays," *Physics Reports*, Vol. 109, pp. 309–426, 1984.

4   K. Wan, A.K. Jain, and J.E. Lukens, "Submillimeter wave generation using Josephson junction arrays," *Appl. Phys. Lett.*, vol. 54, pp. 1805–1807, 1989.

5   K. Wan, B. Bi, A.K. Jain, L.A. Fetter, S. Han, W.H. Mallison and J.E. Lukens, "Refractory submillimeter Josephson effect sources," *IEEE Trans. Magn.*, vol. 27, pp. 3339–3342, 1991.

6   S.P. Benz and C.J. Burroughs, "Coherent emission from two-dimensional Josephson junction arrays," *Appl. Phys. Lett.*, vol. 54, pp. 1805–1807, 1991.

7   M.J. Wengler, A. Pance, B. Liu, and R.E. Miller, "Quasioptical Josephson oscillator," *IEEE Trans. Magn.*, vol. 27, pp. 2708–2711, March, 1991.

8   T. Van Duzer and C.W. Turner, *Principles of Superconductive Devices and Circuits*. New York: Elsevier, 1981.

9   W.C. Stewart, "Current-voltage characteristics of Josephson junctions," *Appl. Phys. Lett.*, vol. 12, pp. 277–280, April, 1968.

10   D.E. McCumber, "Effect of ac Impedance on dc Voltage-Current Characteristics of Superconductor Weak-Link Junctions," *J. Appl. Phys.*, vol. 39, pp. 3113–3118, June, 1968.

11   R.L. Kautz and R. Monaco, "Survey of chaos in the rf-biased Josephson junction," *J. Appl. Phys.*, vol. 57, pp. 875 889, February, 1985.

12   A.D. Smith, R.D. Sandell, A.H. Silver and J.F. Burch, "Chaos and Bifurcation in Josephson Voltage-Controlled Oscillators," *IEEE Trans. Magn.*, vol. 23, pp. 1267–1270, March, 1987.

13   G.S. Lee and S.E. Schwarz, "Numerical and analytical studies of mutual locking of Josephson tunnel junctions," *J. Appl. Phys.*, vol. 55, pp. 1035–1043, February, 1984.

14   R.G. Hicks, P.J. Khan, "Numerical Analysis of Nonlinear Solid-State excitation in Microwave Circuits," *IEEE Trans. Microwave Theory and Techniques*, vol. 30, pp. 251–259, March, 1982.

15   J.R. Tucker and M.J. Feldman, "Quantum detection at millimeter wavelengths," *Rev. Mod. Phys.*, vol. 57, pp. 1055–1113, 1985.

16   W.W. Lam, C.F. Jou, H.Z. Chen, K.S. Stolt, N.C. Luhmann Jr. and D.B. Rutledge, "Millimeter-wave diode-grid phase shifters," *IEEE Trans. on Microwave Theory and Techniques*, vol. MTT–36, pp. 902–907, 1988.

17   C.F. Jou, W.W. Lam, H.Z. Chen, K.S. Stolt, N.C. Luhmann Jr. and D.B. Rutledge, "Millimeter-wave diode-grid frequency doublers," *IEEE Trans. on Microwave Theory and Techniques*, vol. MTT–36, pp. 1507–1514, 1988.

18   Z.B. Popovic, R.M. Weikle, M. Kim, K.A. Potter, and D.B. Rutledge, "Bar-grid oscillators," *IEEE Trans. Microwave Theory and Techniques*, vol. 38, pp. 225–230, March, 1990.

19   D.B. Rutledge, Z.B. Popovic, R.M. Weikle, M. Kim, K.A. Potter, R.C. Compton, and R.A. York, "Quasi-optical power-combining arrays," *1990 IEEE MTT–S International Microwave Symposium Digest*, 1990.

20   R.E. Collin, *Field Theory of Guided Waves*. New York: McGraw-Hill, 1960.

21   J. B. Hacker, R.M. Weikle, II, M. Kim, M. P. DeLisio and D. B. Rutledge, "A 100-element planar Schottky diode grid mixer," *IEEE Trans. on Microwave Theory and Techniques*, submitted, 1991.

22 A. Pance and M.J. Wengler, "Microwave modeling of 2–D active grid antenna arrays," *IEEE Trans. Microwave Theory and Techniques*, submitted, 1991.

23 D.B. Rutledge, D.P. Neikirk and D.P. Kasilingam, "Integrated-Circuit Antennas," In *Infrared and Millimeter waves*, K.J Button, Ed. vol. 10, pp. 1–90, New York: Academic Press, 1983.

24 D. Rogovin and D.J. Scalapino, "Fluctuation phenomena in tunnel junctions," *Ann. Phys.*, Vol. 86, pp. 1–90, July 1974.

25 R.P. Robertazzi, B.D. Hunt and R.A. Buhrman, "Coupled Tunnel Junction Experiments at the Gap Frequency," *IEEE Trans. Magn.*, vol. 23, pp. 1271–1274, 1987.

Thermal Reaction Chemistry of Polycyclic Alkylaromatics Chemical Models of Petroleum Asphaltenes

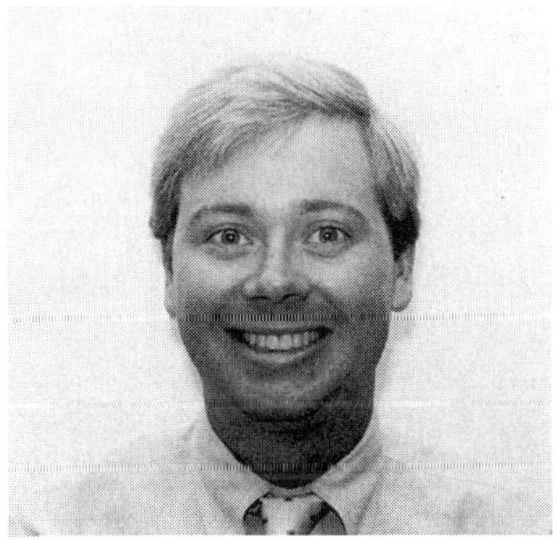

C. MICHAEL SMITH

Department of Chemical Engineering
University of Michigan
Research Supervisor: Philip E. Savage

ABSTRACT

A family of alkyl-substituted polycyclic aromatic hydrocarbons was pyrolyzed in microbatch reactors for holding times ranging from 0–500 minutes at temperatures between 350°C and 425°C. A general pyrolysis network was deduced for these compounds, and it comprised two major and one minor parallel pathways. The first major pathway resulted in products analogous to the major products observed from alkylbenzene pyrolysis. The second major pathway led to products via the cleavage of the strong aryl-alkyl C–C bond. The third pathway led to small amounts of products, presumably through cyclization and condensation reactions. The relative importance of the two major pathways varied for the different compounds. The rates of aryl-alkyl bond cleavage were different for the different compounds, and these rates were quantitatively related to the compounds' localization energies through Dewar reactivity numbers. The resulting correlation provides an excellent tool for predicting the rate of aryl-alkyl bond cleavage for any n-alkylaromatic based solely upon the size and configuration of the aromatic nucleus and the point of substitution of the alkyl chain. Further experimental studies which employed molecular probes and mechanistic modeling studies revealed that a selective mechanism such as molecular disproportionation or radical hydrogen transfer was likely responsible for the observed hydrogenolysis.

INTRODUCTION

Because the reserves of conventional light crudes oils are finite, the industrial trend toward processing heavy crudes will continue. This change in feedstock character requires engineers to more frequently encounter many of the processing difficulties associated with heavy oils such as more rapid catalyst deactivation and coke formation. Petroleum asphaltenes, which are ubiquitous in heavy crudes, are often the key agents responsible for these processing problems. Because asphaltenes are hydrogen deficient, their presence necessitates a processing strategy based either on carbon rejection (e.g., coking) or upgrading through hydrogen addition (e.g., hydrotreating). Since heavy crudes contain appreciable portions (10–30%) of asphaltenes, reaction models that describe these chemical processes are valuable. Often the confidence in deduced reaction pathways and kinetic models increases with the understanding of the fundamental chemistry of the complex reaction system. Thus, our goal was to resolve the reaction fundamentals for asphaltene thermolysis.

One approach for elucidating the chemistry of complex feedstocks, such as asphaltenes, involves the selection of and experimentation with simple compounds that serve as models of important reactive moieties within the complex feedstock. The current understanding of asphaltene microstructure (Mieville et al., 1989; Ali et al., 1990; Speight, 1988; Speight, 1989; Waller et al., 1989; El-Mohamed et al., 1986) indicates that clusters of aromatic rings linked together through aliphatic chains are important structural features. Typical clusters can contain up to six rings, and aliphatic chains can contain from one to 20 carbon atoms. Thus, n-alkylaromatic compounds are relevant chemical models of the hydrocarbon portion of petroleum asphaltenes.

The pyrolysis of n-alkylbenzenes, the simplest compounds in this class, has been extensively investigated (e.g., Mushrush and Hazlett, 1984; Blouri et al., 1985; Savage and Klein, 1987a,b; Freund and Olmstead, 1989). The results of these works showed that the alkylbenzene pyrolysis network involves free-radical chemistry and leads to three product lumps. For example, the pyrolysis pathway for n-pentadecylbenzene (PDB) leads to toluene and 1-tetradecene as a product pair, styrene and tridecane as a second pair, and series of alkylbenzenes, alkenylbenzenes, n-alkanes, and a-olefins in much lower yields as the third product lump (Savage and Klein, 1987a).

Although the pyrolysis of n-alkylbenzenes is now well understood, much less is known about the pyrolysis of polycyclic n-alkylaromatics. Recent pyrolyses of n-alkylpyrenes, however, have revealed that the pyrolysis network for alkylbenzenes does not completely describe the pyrolysis pathways for these polycyclic alkylaromatics (Smith and Savage, 1991a,b; Savage et al., 1989; Freund et al., 1990). The pyrolysis of 1-dodecylpyrene (DDP), for instance, led to three product lumps analogous to those observed from alkylbenzene pyrolysis (e.g., methylpyrene plus 1-undecene, vinylpyrene plus decane, and series of alkylpyrenes, alkenylpyrenes, alkenes, and alkanes) plus an additional major product pair, pyrene and dodecane. These latter two products can be formed only through cleavage of the aryl-alkyl C–C bond, and their presence indicates the existence of a new pyrolysis pathway for 1-dodecylpyrene. The presence of this pathway, which was insignificant in alkylbenzene pyrolyses, was unexpected because the C–C aryl-alkyl bond is the strongest in the alkyl chain, and its bond dissociation energy is about 100 kcal/mol (McMillen and Golden, 1982).

The marked differences between the pyrolysis pathways and kinetics for alkylbenzenes and alkylpyrenes and the relevance of such compounds as chemical models for asphaltenes and heavy crude oils motivated the research described in this paper. During the past year we sought to determine the effects of the structure of the n-alkylaromatic on the reaction pathways and kinetics. The specific variables of interest were (1) the number of aromatic rings, (2) the configuration of the aromatic rings, and (3) the position at which the aliphatic substituent resides upon the aromatic nucleus. Accordingly, we have pyrolyzed a set of 15 different n-alkyl aromatics, and these are displayed in Figure 1. This paper provides the results of our experiments and then addresses their implications in terms of a general pyrolysis network for n-alkylaromatics and a correlation of the compounds' structures and reactivities. Furthermore, we address the discrimination between hydrogenolysis mechanisms through experimental studies that employ the use of molecular probes and the development of a pyrolysis model for 1-methylpyrene from elementary steps.

EXPERIMENTAL

The pyrolyses of the model compounds in Figure 1 were conducted in microbatch reactors neat and in some instances in a benzene diluent. Batch holding times ranged up to 500 minutes, and the reaction temperatures were between 350°C and 425°C.

All of the model compounds in Figure 1 were available in high purity from commercial sources except for 1-n-dodecylnaphthalene, (DDN) which was obtained in 99% purity through a custom synthesis by American Tokyo Kasei. 2-n-Dodecylphenanthrene (DDH), 1-n-undecylnaphthalene (UDN), 9-n-dodecylanthracene (DDA), 6-n-octylchrysene (OC), and 3-n-hexylperylene (HP) were obtained from the American Petroleum Institute Project 42 through the Thermodynamic Research Center at Texas A&M University. 2-n-Butylnaphthalene (BN) was obtained from the American Petroleum Institute Standard Reference Materials through Carnegie Mellon University. 1-n-Dodecylpyrene (DDP) and all other alkylpyrenes were obtained from either Molecular Probes, Texas A&M University, or Carnegie Mellon University. Dodecylbenzene (DDB), biphenyl, and benzene were obtained from Aldrich.

All pyrolyses were conducted in constant-volume, 316 stainless steel, microbatch reactors. The reactors were constructed from one Swagelok port connector and two Swagelok end caps, and they had a nominal volume 0.6 ml. The

Model Compounds	Name	Carbon Chain Length	Dewar Number
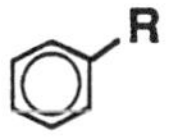	DDB	12	2.31
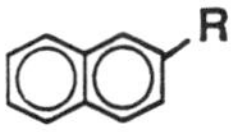	BN	4	2.12
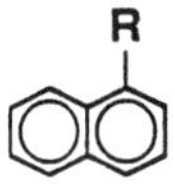	UDN,DDN	11,12	1.81
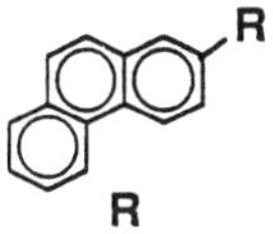	DDH	12	2.18
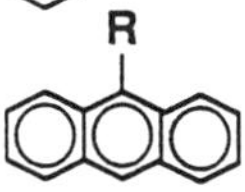	DDA	12	1.26
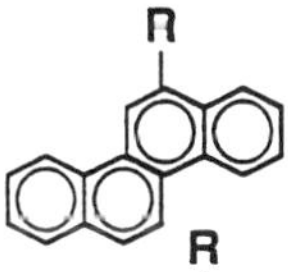	OC	8	1.67
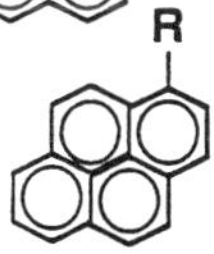	DDP	1-16	1.51
	HP	6	1.33

Figure 1. Petroleum Asphaltene Model Compounds

batch reactors were typically loaded with 10 mg of the model compound and 10 mg of an internal standard (e.g., biphenyl). An inert reaction environment was provided by purging the reactors with argon before closing. After being loaded and closed, the reactors were placed in an isothermal fluidized sand bath at the desired pyrolysis temperature. Upon reaching the desired holding time, the reactors were removed from the fluidized bath, and the reaction was quenched by placing the reactors in an ambient temperature water bath. The reactors were then opened, and the products were recovered by repeated extraction with benzene.

The reaction products were routinely analyzed by gas chromatography (GC) and gas chromatography-mass spectrometry (GC–MS). The GC analysis used a Hewlett Packard (HP) model 5890 instrument equipped with a HP 3392A integrator, a flame ionization detector (FID), and a HP 7673A autosampler. We used both a 5-m x 0.53 mm x 2.55 µm film thickness HP–1 methyl silicone capillary column and a 12-m x 0.12 mm x 0.33 µm film thickness HP–5 (crosslinked 5% phenyl methyl silicone) capillary column for product separation. The GC-MS system included a HP 5890 Series II GC, a HP 5970 mass spectrometric detector, and the HP 59940 MS Chemstation.

RESULTS

In the following section, we provide the results of the neat pyrolyses of UDN, DDH, DDA, and OC. Further pyrolysis results are described by Smith and Savage (1991c). Tables I through IV present representative data in terms of the yields of the major products and selected minor products.

1-Undecylnaphthalene (UDN): The neat pyrolysis of UDN was conducted at 375°C, 400°C, and 425°C for batch holding times up to 400 minutes. Table I provides representative experimental data obtained under a variety of different reaction conditions. The most abundant products at low conversions were 1-methylnaphthalene, 1-ethylnaphthalene, decene, and nonane. Less abundant products and those formed at higher conversions included decane, naphthalene, undecane, tetrahydrophenanthrene, phenanthrene, and series of 1-naphthylalkanes and naphthylolefins with alkyl chains containing 3–10 carbon atoms. Of special interest among these minor products was undecane and naphthalene. These products were not present at the shortest batch holding times, but at long times they became major products. For instance, pyrolysis at 425°C for 170 minutes resulted in molar yields of naphthalene and undecane of 0.086 and 0.045 respectively. Note that the formation of this product pair requires the cleavage of the strong aryl-alkyl C–C bond in UDN.

Figure 2 displays the temporal variations of the major low-conversion products for UDN pyrolysis at 425°C. The molar yields of decene and 1-methyl-naphthalene were approximately equal at short batch holding times. For example, at a holding time of 10 minutes the molar yields of 1-methylnaph-

Table I: Product Molar Yields from 1-Undecylnaphthalene Pyrolysis

Pyrolysis Temperature (°C)	375	375	375	375	400	400	400	400	425	425	425	425
Time (min)	20	200	300	400	20	60	145	190	10	60	90	170
Nonane	0.040	0.034	0.050	0.073	0.005	0.017	0.006	0.048	0.010	0.056	0.065	0.087
Decene	0.035	0.082	0.098	0.11	0.013	0.035	0.038	0.038	0.024	0.049	0.035	0.011
Decane	0.005	0.013	0.022	0.039	0.000	0.003	0.016	0.027	0.000	0.022	0.034	0.063
Undecane	0.000	0.013	0.024	0.043	0.000	0.004	0.014	0.019	0.000	0.015	0.024	0.045
Naphthalene	0.000	0.017	0.029	0.053	0.000	0.005	0.016	0.022	0.000	0.020	0.037	0.086
Methylnaphthalene	0.045	0.13	0.19	0.25	0.018	0.054	0.11	0.14	0.034	0.17	0.22	0.28
Ethylnaphthalene	0.006	0.031	0.047	0.071	0.003	0.017	0.039	0.047	0.007	0.073	0.094	0.12
Vinylnaphthalene	0.000	0.000	0.000	0.000	0.002	0.000	0.000	0.000	0.006	0.000	0.000	0.000
Undecylnaphthalene	1.0	0.90	0.87	0.83	1.0	0.84	0.69	0.56	0.92	0.50	0.33	0.13

Table II: Product Molar Yields from 2-Dodecylphenanthrene Pyrolysis

Pyrolysis Temperature (°C)	375	375	375	375	400	400	400	400	425	425	425	425
Time (min)	10	163	187	210	30	45	95	223	10	30	50	159
Decene	0.000	0.004	0.004	0.004	0.003	0.005	0.008	0.009	0.006	0.010	0.013	0.005
Decane	0.002	0.034	0.030	0.034	0.021	0.029	0.058	0.088	0.036	0.065	0.096	0.11
Undecene	0.005	0.058	0.058	0.049	0.037	0.047	0.081	0.064	0.058	0.087	0.097	0.018
Undecane	0.002	0.006	0.005	0.004	0.005	0.004	0.007	0.029	0.003	0.006	0.014	0.029
Phenanthrene	0.000	0.004	0.003	0.002	0.004	0.005	0.007	0.004	0.004	0.009	0.010	0.011
Methylphenanthrene	0.006	0.075	0.080	0.059	0.043	0.061	0.116	0.151	0.018	0.14	0.21	0.30
Ethylphenanthrene	0.000	0.026	0.030	0.020	0.016	0.026	0.046	0.087	0.000	0.073	0.11	0.16
Vinylphenanthrene	0.001	0.001	0.000	0.002	0.002	0.002	0.003	0.000	0.010	0.003	0.002	0.004
Dodecylphenanthrene	0.77	0.89	0.86	0.53	0.63	0.63	0.47	0.31	0.67	0.33	0.20	0.02

Table III: Product Molar Yields fro, 9-Dodecylanthrancene Pyrolysis

Pyrolysis Temperature (°C)	350	350	350	350	375	375	375	375	400	400	400	400
Time (min)	10	30	90	180	10	20	30	60	10	15	20	40
Decane	0.000	0.000	0.000	0.002	0.000	0.000	0.002	0.006	0.003	0.005	0.009	0.022
Undecene	0.000	0.000	0.000	0.002	0.001	0.000	0.003	0.006	0.006	0.009	0.012	0.000
Undecane	0.000	0.000	0.004	0.011	0.001	0.003	0.005	0.023	0.012	0.015	0.030	0.050
Dodecene	0.000	0.002	0.009	0.019	0.001	0.003	0.010	0.030	0.012	0.017	0.029	0.031
Dodecane	0.005	0.023	0.096	0.25	0.025	0.044	0.11	0.02	0.12	0.19	0.29	0.58
Dihydroanthracene	0.000	0.002	0.003	0.020	0.000	0.003	0.007	0.020	0.003	0.009	0.011	0.079
Anthracene	0.01	0.05	0.17	0.34	0.063	0.10	0.25	0.45	0.19	0.31	0.39	0.71
Methylanthracene	0.000	0.000	0.000	0.000	0.000	0.000	0.000	0.007	0.000	0.001	0.000	0.003
Dodecylanthrancene	0.71	0.64	0.55	0.04	0.72	0.53	0.26	0.10	0.48	0.28	0.25	0.04

Table IV: Product Molar Yields from 6-Octylchrysene Pyrolysis

Pyrolysis Temperature (°C)	375	375	375	375	400	400	400	400	425	425	425	425
Time (min)	45	215	245	330	15	33	130	180	15	30	45	60
Chrysene	0.000	0.071	0.082	0.074	0.000	0.025	0.13	0.28	0.038	0.12	0.18	0.19
Methylchrysene	0.011	0.047	0.046	0.040	0.016	0.035	0.089	0.12	0.072	0.14	0.17	0.14
Ethylchrysene	0.000	0.020	0.019	0.000	0.000	0.000	0.039	0.039	0.039	0.072	0.076	0.053
Octylchrysene	0.84	0.99	0.79	0.42	0.92	0.90	0.51	0.33	0.92	0.76	0.48	0.19

thalene and decene were 0.034 and 0.024, respectively. The molar yield of decene reached a maximum value of 0.059 at a batch holding time of 45 minutes, and then decreased until it reached its ultimate value of 0.011 at 170 minutes. The molar yield of 1-methylnaphthalene, on the other hand, increased steadily throughout the reaction. It achieved an ultimate value of 0.28 at 170 minutes.

Figure 2 also displays the molar yields of nonane and 1-ethylnaphthalene from the pyrolysis of UDN at 425°C. Similar to the behavior of the yields of 1-methylnaphthalene and decene, the molar yields of nonane and 1-ethylnaphthalene were approximately equal at short reaction times. For example, at 10 minutes the molar yields of nonane and 1-ethylnaphthalene were 0.01 and 0.007 respectively. Both yields also increased with time.

2-Dodecylphenanthrene (DDH): The pyrolysis of 2-dodecylphenanthrene (DDH) was conducted at 375, 400, and 425°C for batch holding times ranging up to 223 minutes, and Table II provides representative results. The major products from DDH pyrolysis were methylphenanthrene, ethylphenanthrene, undecene, and decane. Products formed in lower yields included undecane, 2-vinylphenanthrene, decene, phenanthrene, chrysene, and a series of alkylphenanthrenes. Of the most interest among these minor products was the formation of phenanthrene. The highest molar yield of phenanthrene was 0.011, which was obtained from pyrolysis at 425°C and 159 minutes where the conversion of DDH was 0.98. The molar yields of phenanthrene were always much lower than the yields of the major products.

At low batch holding times the molar yields of undecene and methylphenanthrene were approximately equal. For example at 30 minutes for pyrolysis of DDH at 400°C, the molar yields of methylphenanthrene and undecene were 0.043 and 0.037, respectively. At longer batch holding times, however, the molar yield of undecene reached a maximum value and then decreased whereas the yield of methylphenanthrene increased steadily throughout the reaction. The yields of ethylphenanthrene and decane increased with time, and the yield of decane was always higher than the yield of ethylphenanthrene. For example at 400°C, the molar yields of decane and ethylphenanthrene were 0.021 and 0.016, respectively at a batch holding time of 30 minutes.

9-Dodecylanthracene (DDA): The neat pyrolysis of DDA was conducted at 350, 375 and 400°C, and Table III provides representative results. The major products from DDA pyrolysis under all conditions studied were anthracene and dodecane. The minor products included decane, decene, undecane, undecene, dodecene, methylanthracene, dihydroanthracene, and tetrahydroanthracene. Of these minor products, dodecene, undecane, and dihydroanthracene had the highest molar yields.

Figure 3 presents the temporal variations of the molar yields of the two major products from the neat pyrolysis of DDA at 400°C. The yields of both anthracene and dodecane increased steadily with time, and the anthracene

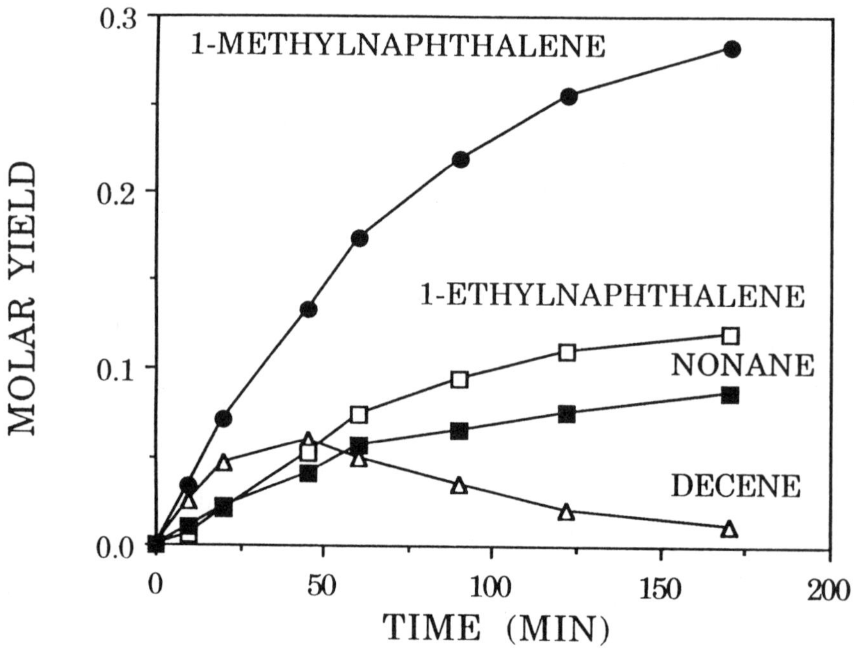

Figure 2. Temporal Variation of Product Molar Yields From UDN Neat Pyrolysis at 425°C.

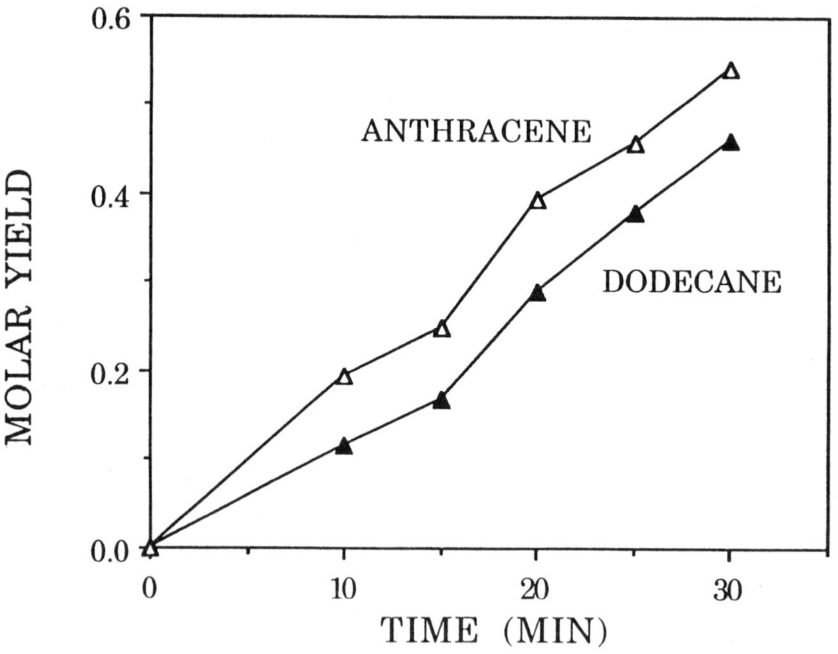

Figure 3. Temporal Variation of Product Molar Yields From DDA Neat Pyrolysis at 400°C.

yield always exceeded the n-dodecane yield. The ultimate yields at 40 minutes were 0.71 for anthracene and 0.58 for n-dodecane.

6-Octylchrysene (OC): The pyrolysis of OC was conducted at 375, 400, and 425°C for batch holding times ranging up to 330 minutes, and Table IV lists representative results.   The major products were chrysene, methylchrysene, and ethylchrysene. No alkanes or olefins were quantified because these products were either gases, or they co-eluted with the solvent peak during GC analysis. At short batch holding times the principal product was methylchrysene, and ethylchrysene and chrysene were present in lower yields. As the holding time increased, however, the yield of chrysene increased much more rapidly than did the yields of methylchrysene or ethylchrysene.

DISCUSSION

Our discussion of the foregoing experimental results will center on three major areas. First, we provide a general reaction network for the pyrolysis of alkylarenes that is consistent with all of our experimental results. Secondly, we use perturbation molecular orbital theory to correlate the structures of the compounds with their apparent reactivities.   Finally, we discuss the possible mechanistic scenarios possible for engendering hydrogenolysis.

General Pyrolysis Pathways: We used two related methodologies to deduce a general reaction network for the pyrolysis of polycyclic n-alkylaromatic compounds. Discrimination between primary and secondary products was achieved through (1) examination of the variation of the products' molar yields with reactant conversion and (2) examination of the variation of the products' selectivities with reactant conversion. We used the former approach in our previous alkylarene pyrolysis studies (Smith and Savage, 1991a,b; Savage et al., 1989), and application of the latter method is illustrated here. This method, termed the Delplot Technique (Bhore et al., 1990) permits pathway demarcation through the analysis of the y-intercepts of selectivity versus conversion plots where the selectivity is calculated as the molar yield divided by the reactant conversion. Products that possess non-zero intercepts are of first rank, and those that possess zero intercepts are of higher rank. The rank is the order of appearance of a product within the reaction network (e.g., products of first rank are primary products) and represents the number of "slow" reactions which lead to the formation of a product.

To illustrate the application of this method we provide Figure 4, which displays Delplots for the major products of UDN pyrolysis at 425°C. The curves in Figure 4 illustrate the trends in the data and are shown only to aid the reader. Inspection of Figure 4a reveals that 1-methylnaphthalene and decene are products of first rank, since extrapolation of their selectivities to zero conversion yields a non-zero intercept. Furthermore, the values of the y-inter-

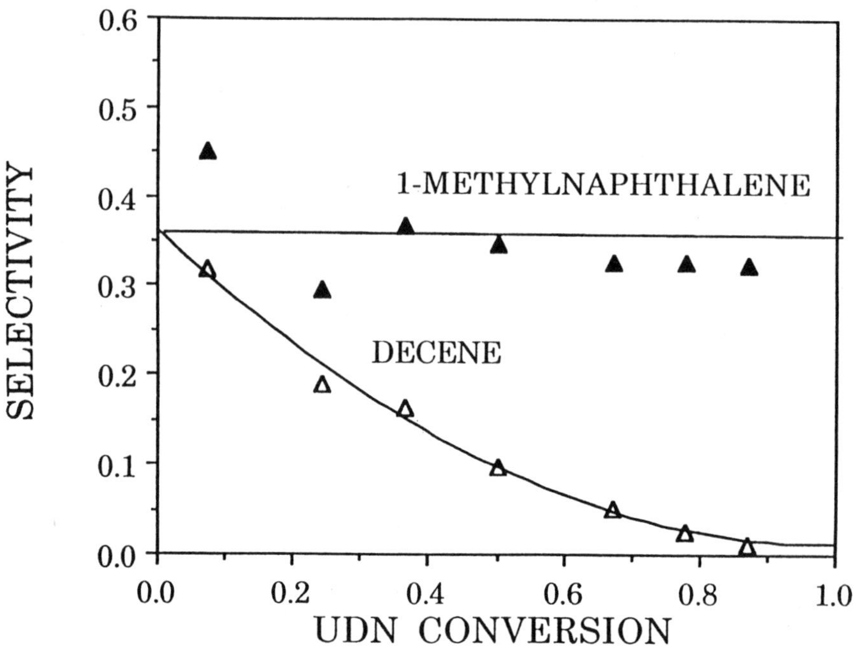

Figure 4a. UDN Selectivity to Major Products (1-Methylnaphthalene and Decene).

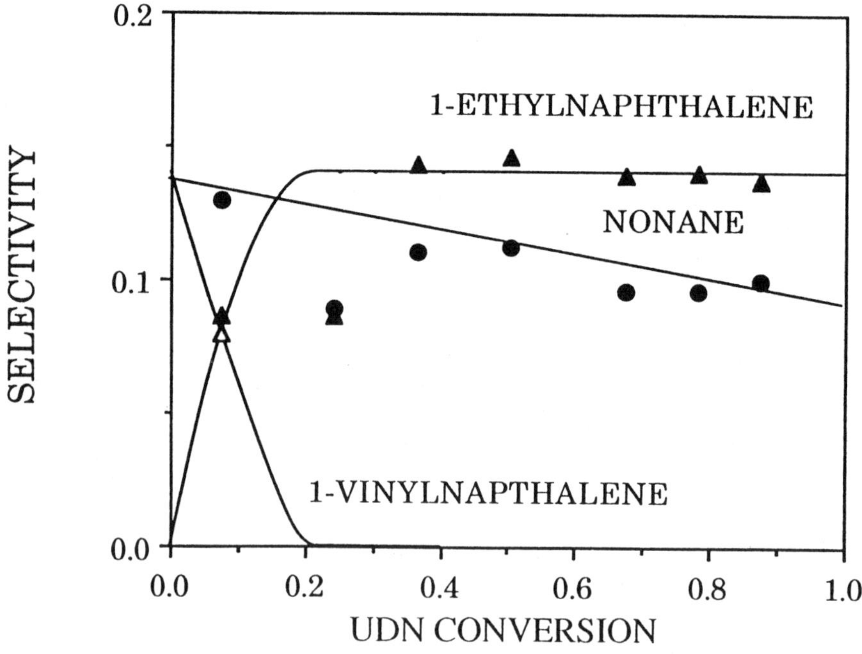

Figure 4b. UDN Selectivity to Major Products (1-Vinylnaphthalene, 1-Ethylnaphthalene and Nonane).

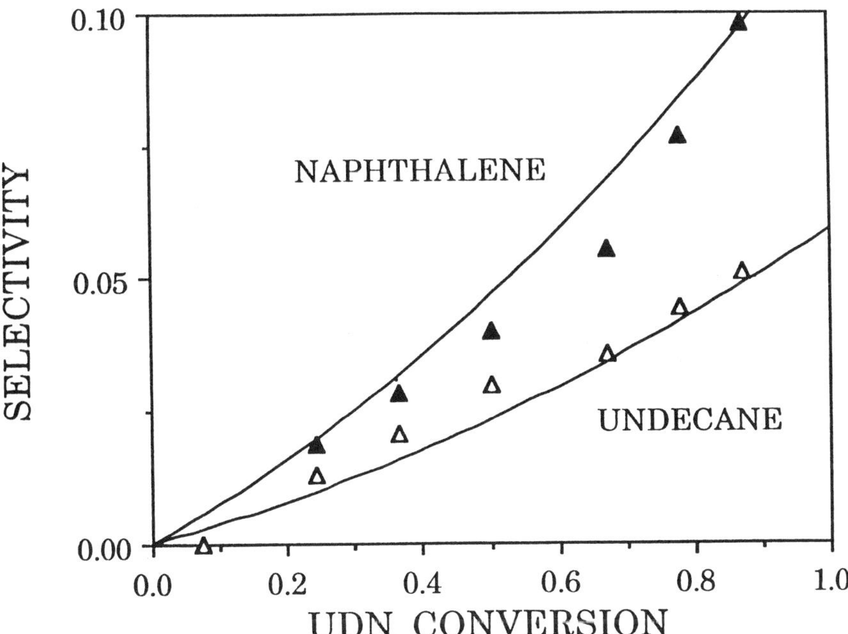

Figure 4c. UDN Selectivity to Major Products (Naphthalene and Dodecane).

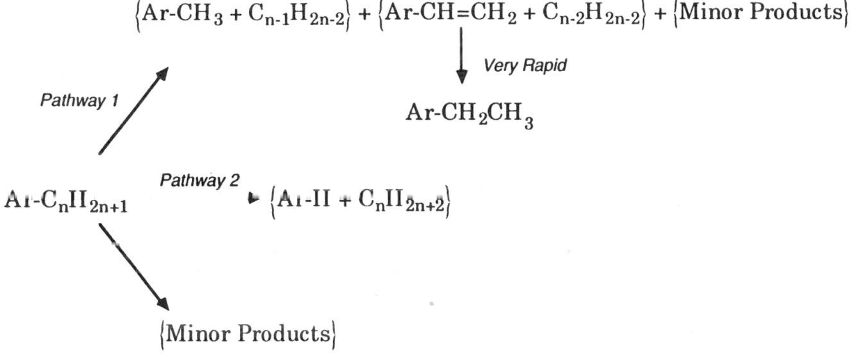

Figure 5. General Pyrolysis Network for Alkyl-Substituted Aromatic Hydrocarbons.

cepts for decene and 1-methylnaphthalene were essentially equal, which suggests that these products were formed in the same reaction step. The selectivity to 1-methylnaphthalene was approximately constant, but the selectivity to decene decreased with conversion suggesting that decene underwent secondary reactions.

Figure 4b is the Delplot for the products vinylnaphthalene, ethylnaphthalene, and nonane. The non-zero and nearly equal intercepts for vinylnaphthalene and nonane reveal that these were products of first rank and that they were formed in the same reaction step. The selectivity to vinylnaphthalene decreased rapidly with conversion while the selectivity to ethylnaphthalene increased rapidly. This behavior is consistent with vinylnaphthalene undergoing secondary reaction to produce ethylnaphthalene. The rapid conversion of styrene to ethylbenzene has been observed by previous investigators (Klein and Virk, 1983; Savage and Klein, 1987a), and we expect this conversion to be even faster for polycyclic vinylarenes (Church and Gleicher, 1976).

Figure 4c displays the selectivities to naphthalene and undecane as a function of UDN conversion. The y-intercepts for these two products were approximately zero, suggesting that naphthalene and undecane are products of rank greater than one. That is, they are not formed during the initial stages of the reaction. Note, however, that this does not necessarily require that these be secondary products. Indeed, the product pair naphthalene and undecane can only be formed via the cleavage of the aryl-alkyl C–C bond in UDN. Thus, these are strictly primary reaction products. The zero initial selectivity and the steady increase in selectivity with UDN conversion are indicative of the formation of naphthalene and undecane through a primary autocatalytic pathway rather than through a secondary pathway.

We conducted similar reaction pathway analyses using the pyrolysis data for the other compounds, and the resulting pathways were completely analogous to those obtained from UDN pyrolysis. Therefore, the foregoing analysis has provided a general reaction network for the pyrolysis of alkylarenes. This pyrolysis network, shown in Figure 5, comprises three parallel primary pathways. The first pathway leads to the methylarene plus the C_{n-1} alkene as one product pair (e.g., 1-methylnaphthalene and 1-decene for UDN pyrolysis), the vinylarene and the C_{n-2} alkane as a second product pair (e.g., 1-vinylnaphthalene and nonane for UDN pyrolysis), and series of a-olefins, alkanes, alkylaromatics, and alkenylaromatics as a third grouping involving minor products. The second primary pathway for alkylarene pyrolysis leads to the arene and the C_n alkane (e.g., naphthalene and n-undecane for UDN pyrolysis). These products arise from the cleavage of the aryl-alkyl C–C bond. The final pathway allows for the formation of additional minor products such as those arising from condensation and cyclization reactions. In addition to displaying the three parallel primary reactions, Figure 5 also displays the secondary conversion of vinylarene to, among other products, ethylarene. This reaction is very rapid. Other secondary reactions also occur (e.g., decomposition of alkenes, dealkylation of methyl- and ethylaromatics), but these are not shown explicitly in Figure 5.

In summary, our experimental results are consistent with the pyrolysis network shown in Figure 5, and with the identity of the aromatic moiety and the position of the substitutent being key variables in determining the relative importance of pathways 1 and 2. Pathway 1 was the most important in guiding the pyrolyses of DDB, BN, DDH, and UDN. Both pathways 1 and 2 were important for OC and DDP, but pathway 2 was the dominant route for the pyrolyses of DDA and HP.

Structure and Reactivity: The relative importance of pathways 1 and 2 was dependent upon the structure of the n-alkylaromatic. This finding indicates that the structure of the compound influenced its reactivity.

Our discussion of the effect of the structure of the aromatic portion of the reactants on their reactivity will focus on pathway 2, which involves cleavage of the aryl-alkyl C–C bond. The mechanism responsible for this pathway involves a hydrogenolysis reaction in which the alkyl substituent is displaced by hydrogen (Smith and Savage, 1991a,b), but the mechanistic details remain unresolved. Possible mechanisms include hydrogen atom ipso-substitution (Vernon, 1980), molecular disproportionation (Billmers et al., 1986, 1989), and radical hydrogen transfer (Malhotra and McMillen, 1990; McMillen et al., 1987). All of these mechanisms involve the transfer of a hydrogen atom from a donor to the *ipso* position of the alkylaromatic thereby engendering aryl-alkyl bond cleavage. Thus, in essence, pathway 2 involves a substitution reaction in which the alkyl chain is replaced by a hydrogen atom. Further discussion concerning hydrogenolysis mechanisms will be given in the following section.

For a family of aromatic substitution reactions, the change in energy of reaction can be attributed to the change in energy of the delocalized electrons (π-energy of the system, ΔE_π) provided there is little or no change in energy of the localized bonds and in solvation (Dewar and Dougherty, 1975; Dewar 1969; Streitwieser, 1961). ΔE_π is a measure of the difference in the p-electron energy of the reactant aromatic system and the π-electron energy of the "Wheland intermediate" or σ-complex and it is often termed the localization energy (Streitwieser, 1961). Invoking the Evans-Polanyi relationship (Boudart, 1968; Dewar and Dougherty, 1975; Dewar, 1969) then allows us to relate the change in activation energy, ΔE_a, between members of the reaction family with their corresponding change in ΔE_π. This result is given as Equation 1 where a is the Evans-Polanyi factor.

$$\Delta E_a = constant + \alpha\, \Delta E_\pi \qquad\qquad (1)$$

For even alternant hydrocarbons, ΔE_π can be calculated readily from perturbation molecular orbital theory (Dewar and Dougherty, 1975; Dewar 1969; Streitwieser, 1961) as

$$\Delta E_p = 2\,\beta\,(a_{or} + a_{os}) = b\,N_t \qquad\qquad (2)$$

where β is the resonance integral, and a_{or} and a_{os} are the coefficients of the non-bonding molecular orbitals at the positions adjacent to the position of substitution, which is denoted by the subscript, t. The quantity $2(a_{or} + a_{os})$ is defined as the Dewar reactivity number, N_t. Dewar (1952) and Gore (1954) provide values of this reactivity number for a large number of different compounds, and the values relevant to the present study were given in Figure 1 and Table VII.

Combining Equations 1 and 2 shows that

$$\Delta E_a = constant + \alpha\,\beta\,N_t \qquad\qquad (3)$$

Equation 3 suggests that a semilog plot of the reaction rates for aromatic substitution reactions as a function of the Dewar reactivity number, N_t, should be linear. The literature presents many such correlations of reactivity in aromatic substitution reactions with a measure of the localization energy. (e.g., Altshuler and Berliner, 1966; Dickerman et al., 1973; Dewar and Thompson, 1965).

To develop a correlation using the present results, we calculated the initial rate of aryl-alkyl bond cleavage for each of the alkylaromatics studied. These rates were calculated as the initial slope of the molar yield vs. time curve for both the arene and, when possible, for the corresponding n-alkane. Figure 6 displays the natural logarithm of the initial rate of aryl-alkyl bond cleavage as a function of the Dewar reactivity number for the alkylarenes used in this study and for compounds pyrolyzed previously (Savage et al., 1989; Savage and Klein, 1987b). The open symbols represent those rates calculated from the alkane, and the filled symbols represent rates calculated from the arene. The correlation of the initial rates with the Dewar reactivity number is clearly linear for all three temperatures.

Figure 6 is useful, not only because it summarizes a large set of experimental kinetics data, but also because it possesses predictive capabilities. That is, the initial rate and relative importance of aryl-alkyl cleavage for any n-alkylaromatic can be predicted *a priori* from Figure 6. For instance, a 1-alkylanthracene, which has a Dewar number of 1.57, is expected to undergo facile aryl-alkyl cleavage, whereas a 2-alkyltriphenylene, which has a much higher Dewar number of 2.12 will undergo very little aryl-alkyl cleavage. Its pyrolysis will follow pathway 1 predominantly. To summarize, we have successfully correlated the rate of aryl-alkyl bond cleavage for the pyrolysis of alkyl-substituted aromatic compounds with the structure of the aromatic moiety using results from perturbation molecular orbital theory.

As one additional correlation of structure and reactivity, consider Figure 7. Here we use semilog coordinates to plot the selectivities toward pathways 1 and 2 versus the Dewar reactivity number. The selectivity toward pathway 1,

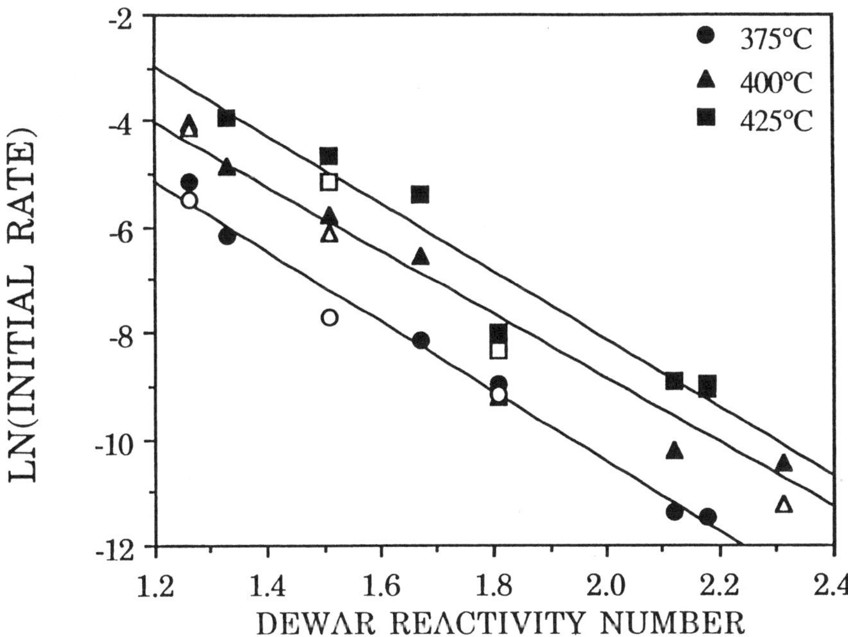

Figure 6. Correlation of Initial Rate of Aryl-Alkyl Bond Cleavage in Alkylarene Pyrolysis with the Dewar Reactivity Numbers.

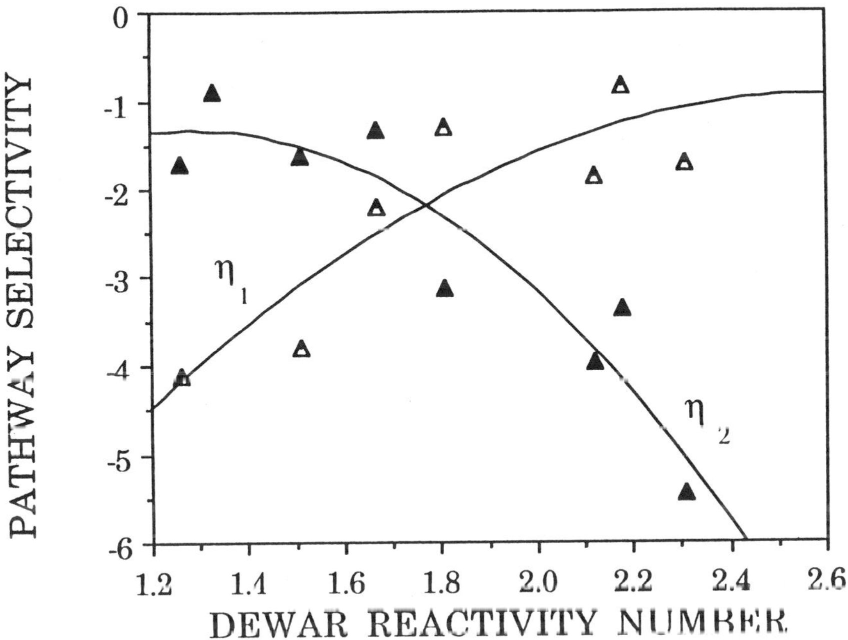

Figure 7. Correlation of Pathway Selectivity with Dewar Reactivity Number for Alkylarene Pyrolysis at 400°C.

η_1, was calculated as the ratio of the initial rate of methylaromatic formation to the pseudo-first order rate constant for alkylarene disappearance. Likewise, the selectivity toward pathway 2, η_2, was calculated as the ratio of the initial rate of arene formation to the pseudo-first order rate constant for alkylarene disappearance. The results show that η_2 is high and essentially Dewar-number invariant for compounds with low Dewar numbers (e.g., <1.6). As the Dewar number increases, however, η_2 decreases in a nearly linear fashion. Conversely, h_1 increases as the Dewar number increases, and this behavior indicates an increase in the importance of pathway 1. Thus, as the Dewar number changes in Figure 7, the relative importance of the two pathways shifts. These results are entirely consistent with our previous assertion that n-alkylarenes fall into one of three categories. Compounds with low Dewar numbers pyrolyze predominantly via pathway 2. Compounds with high Dewar numbers pyrolyze predominantly via pathway 1. Compounds with intermediate Dewar numbers experience contributions from both pathways 1 and 2.

Finally, it is interesting to note that the shape of the curve for η_2 is reminiscent of that for the effectiveness factor as a function of the Thiele modulus. Such a curve arises in heterogeneous catalysis for largely the same types of reasons. There are two rate processes (diffusion and reaction), and the relative importance of the processes shifts with the value of the Thiele modulus.

Hydrogenolysis Mechanisms: The aryl-alkyl C–C bond cleavage observed during DDP pyrolysis is the major pathway at high concentrations and conversions. Although there is currently no unequivocal mechanistic explanation for this hydrogenolysis pathway, the literature provides several possibilities. Figure 8 offers three such mechanisms of hydrogen transfer that can engender strong bond cleavage. These are hydrogen atom ipso-substitution, radical hydrogen transfer (RHT), and molecular disproportionation (MD). There are, of course, other candidates for hydrogen transfer that lead to cleavage of strong carbon-carbon bonds (e.g., 3-step hydrogen transfer). All of these mechanisms involve the addition of hydrogen to the ipso-position of DDP and subsequent elimination of the n-dodecyl substituent. The essential differences between the three mechanisms in Figure 8 are the species that donate the hydrogen. In H-atom ipso substitution (Vernon, 1980) the H-donor is atomic hydrogen. In the second mechanism, RHT (McMillen et al. 1987), a free radical donates hydrogen to the ipso position in the DDP molecule, which can then further react to yield pyrene and the n-dodecyl radical. Potential radical hydrogen donors (RHT in Figure 8) include cyclohexadienyl, allylic, and aliphatic radicals. The third mechanism, MD (Billmers et al., 1986), which is the reverse of radical disproportionation, involves the transfer of hydrogen from a saturated carbon atom to an unsaturated carbon atom and results in the formation of two free radicals. A weakly bound hydrogen atom, such as one attached to the a carbon in DDP, is a possible candidate for being transferred by MD.

Radical hydrogen transfer and molecular disproportionation are reportedly (Malhotra and McMillen, 1990) more selective processes than hydrogen atom

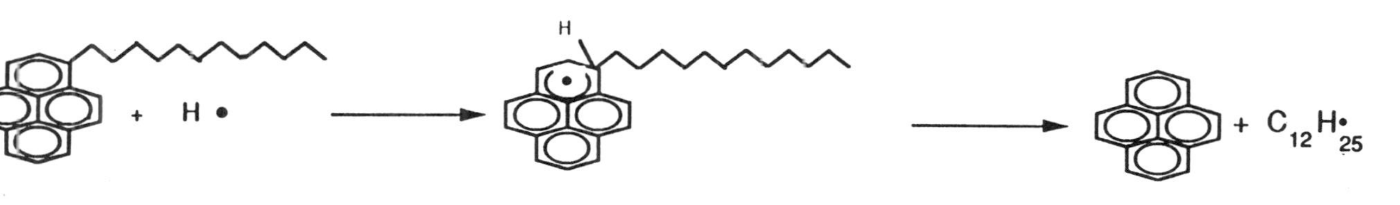
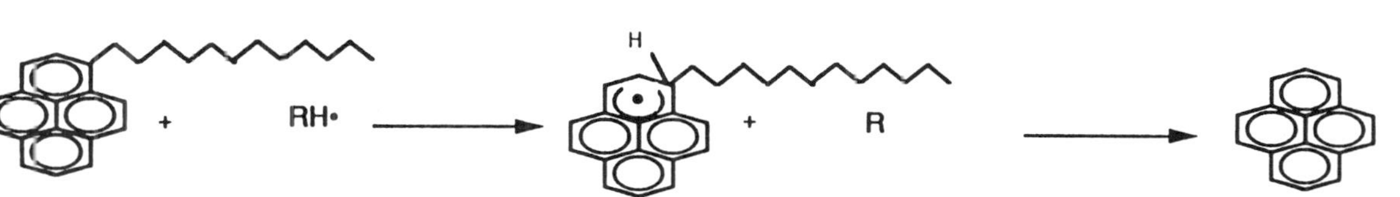
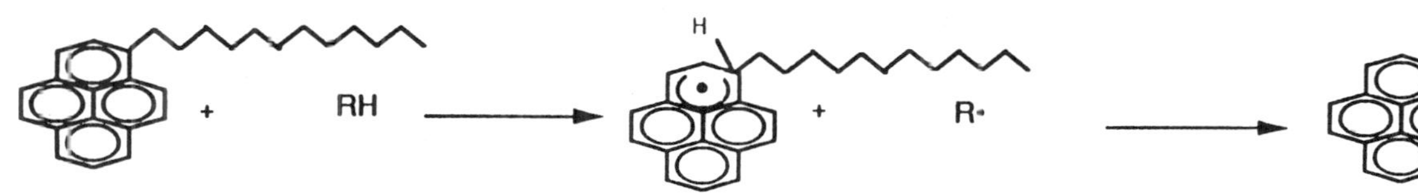

Figure 8. Hydrogenolysis Mechanisms.

addition, thus experimental probes that exploit this difference in selectivity can be used to discriminate between H-atom addition and the more selective hydrogenolysis mechanisms (McMillen et al., 1987; Bockrath et al., 1989). For example, one can use 1,6-dimethylnaphthalene as a reporter molecule to probe the reaction mechanism for DDP pyrolysis. RHT and MD are expected to cleave preferentially methyl groups on the 1-position in 1,6-dimethylnaphthalene because of the greater resonance stabilization energy associated with this position. H-atom addition, on the other hand, is expected to attack both the 1 and 6 positions at roughly equal rates. Thus, measuring the relative yields of the cleavage products, 2-methylnaphthalene and 1-methylnaphthalene, provides a means of assessing the relative importance of selective and non-selective hydrogenolysis mechanisms. McMillen et al. (1987) used thermochemical arguments to predict that the hydrogen atom addition mechanism should yield a molar ratio of 2-methylnaphthalene/1-methylnaphthalene of 1.3–1.6. Corroboration of this estimate can be found by extrapolating rate laws for the hydrodealkylation of 1- and 2-methylnaphthalenes reported by Bixel et al. (1964). We used their data to calculate that the 2-methylnaphthalene/1-methylnaphthalene ratio should range between 1 and 1.7. This range is in good accord with the McMillen et al. (1987) estimate provided that H-atom addition is the dominant mechanism during the hydrodealkylation experiments of Bixel et al. (1964).

To verify the numbers given above, which were obtained using estimations and extrapolations, we conducted hydropyrolysis experiments with 1,6-dimethylnaphthalene at 1500 psig H_2 for a batch holding time of 15 minutes at 400C. These experiments resulted in a ratio of 2-methylnaphthalene/1-methylnaphthalene equal to 2.8 ± 0.5 at the 95% confidence interval. One possible explanation for this experimentally determined ratio being higher than the estimated and extrapolated ratios is that H-atom addition was not the sole hydrogenolysis mechanism operative during the hydropyrolysis experiments. Although the high hydrogen pressure should provide ample numbers of H atoms, hydrogenated 1,6-dimethylnaphthalenes were the major hydropyrolysis products, and these compounds could engender hydrogenolysis by RHT of MD. In any event, the foregoing discussion shows that if RHT or MD is the major mechanism, the resulting 2-methylnaphthlene/1-methylnaphthalene ratio should be greater than 2.8, our experimental estimate of the upper bound for this ratio from H-atom ipso substitution.

In the present work, we reacted DDP with 1,6-dimethylnaphthalene, the hydrogen transfer "reporter" molecule, for 150 minutes at 400°C with the molar ratio of 1,6-dimethylnaphthalene to DDP being about 50. The average ratio of the yields of 2-methylnaphthalene to 1-methylnaphthalene was 4.7 ± 0.4 at the 95% confidence interval. In order to test for statistical significance between these two means (e.g., 4.7 and 2.8) we conducted a two-sample t test and determined that the hypothesis $\mu_{Hydropyrolysis} = \mu_{DDP\ Pyrolysis}$ could be safely rejected at the 0.001 level. This analysis indicates that the cleavage is not due solely to the addition of free hydrogen atoms, but that another more selective mechanism must be operative. RHT and MD are likely candidates.

Further mechanistic insight can be derived from the slopes of the lines in from Figure 6 which are equal to $\alpha\beta$ /RT where R is the gas constant and T is the pyrolysis temperature. The quantity $\alpha\beta$, sometimes termed β_x, was calculated to be 8.6 ± 1.2 kcal/mol at the 95% confidence interval. This quantity provides insight to the mechanism of the substitution reaction and the relative position of the transition state along the reaction coordinate because it is a measure of how closely the transition state resembles the σ-complex. High values of β_x imply late transition states and selective reactions whereas low values of β_x imply early transition states and less selective substitution reactions.

To place the present value of β_x = 8.6 kcal/mole within the context of other work with aromatic substitution reactions, we note that β_x = 3.4 for phenylation, 4.7 for nitration, 5.5 for methylation, 7.5 for deuteriodeprotonation, 14.5 for chlorination, and 16.1 for bromination (Altschuler and Berliner, 1966; Dickerman et al., 1973) Thus, bromination has the latest transition state and phenylation has the earliest transition state. The magnitude of β_x obtained from our analysis indicates that the transition state occupies an intermediate position along the reaction coordinate between the highly selective chlorination reaction and the less selective deuteriodeprotonation reaction. Therefore, we conclude that the mechanism responsible for engendering the aryl-alkyl C–C bond cleavage is moderately selective. This conclusion is consistent with our previous experimental results employing 1,6 dimethylnaphthalene as a molecular probe.

To further explore the possible mechanistic scenarios, we devleoped a mechanistic model for a simple alkylarene, 1-methylpyrene. Our development of mechanistic models for methylpyrene pyrolysis was guided by our experimental observations, by previous mechanistic models for toluene pyrolysis, and by previous investigations into the pyrolysis of polycyclic aromatic hydrocarbons. The 19 elementary step reaction mechanism used to describe the pyrolysis of methylpyrene is depicted in Figure 9, and the Arrhenius parameters estimated for each step are listed in Table V. Methylpyrene undergoes initiation through two possible routes: unimolecular homolytic dissociation and bimolecular MD. Chain propagation proceeds via RHT, H-atom and methyl radical addition, and methyl radical and H-atom elimination. In this chain mechanism, H-atoms, methyl radicals, and methylhydropyrenyl radicals act as the chain carriers. Termination occurs through the recombination of methylpyrenyl radicals, which are generated during initiation and by hydrogen abstraction from methylpyrene. Further details are discussed by Smith and Savage (1991d).

We simulated the pyrolysis of ethyl and methylpyrene using Accuchem, a software program developed by the National Institute of Standards and Technology (Braun et al, 1988). The program calculates reactant and product concentrations as a function of time for reactions in a constant volume batch

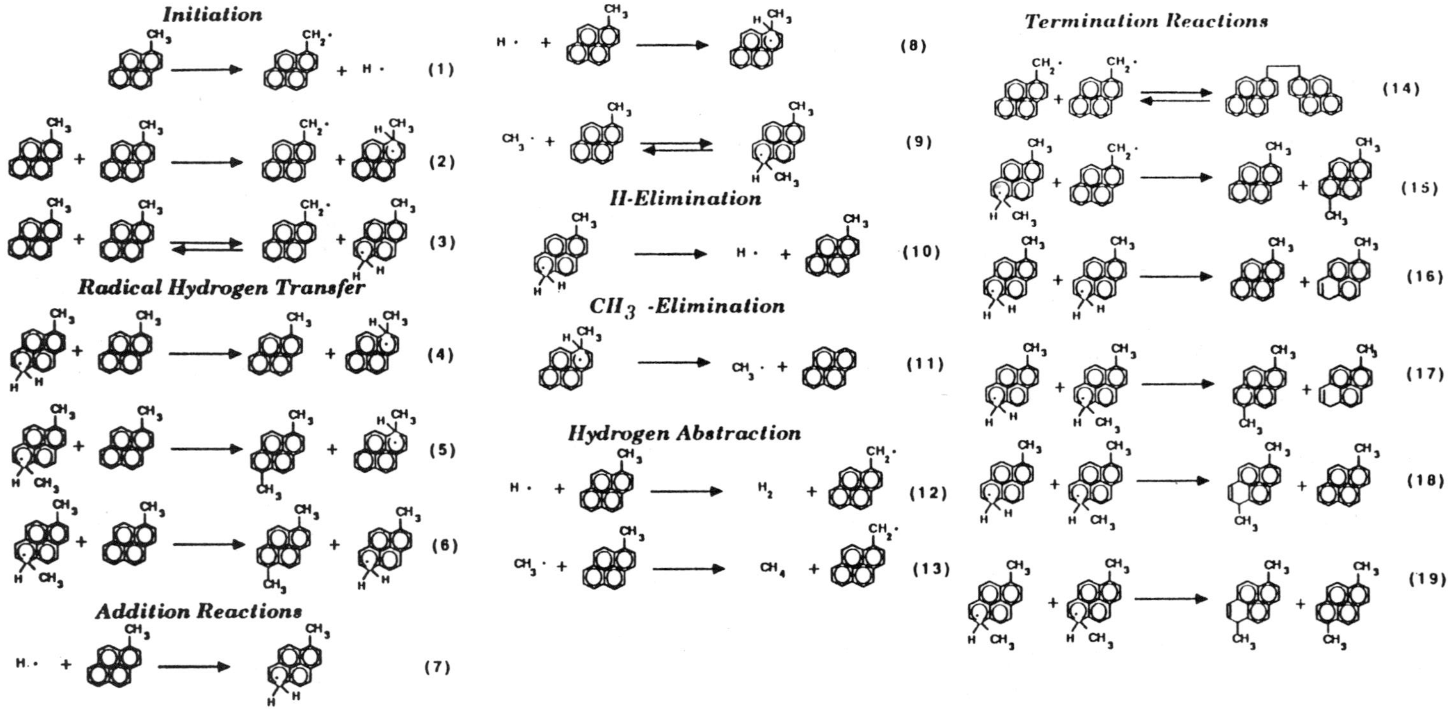

Figure 9. Methylpyrene Pyrolysis Mechanism.

Table V: Arrhenius Parameters used for the simulation of Methylpyrene Pyrolysis

Reaction Number	$\log_{10} A$ (sec^{-1} or Liter mole^{-1} sec^{-1})	Activation Energy (kcal mole^{-1})	Reaction Path Degeneracy
1	16	82.9	3
2	8	45.8	6
3	8	45.8	18
-3	9.5	0	1
4	8.1	16.5	2
5	8.1	16.5	1
6	8.1	16.5	3
7	10.4	2.3	3
8	10.4	2.3	1
9	8.8	4.1	3
-9	13.9	29.4	1
10	13.9	35.3	2
11	13.9	29.4	1
12	11.1	5.6	3
13	8.5	6.9	3
14	9.5	0.0	1
-14	16	51.2	1
15	9.5	0	1
16	9.5	0	1
17	9.5	0	1
18	9.5	0	1
19	9.5	0	1

reactor. Figure 10 displays the comparison of the model prediction and experimental observation for methylpyrene pyrolysis at 400, 425, and 450°C. The model (solid lines) simulates the experimental data (discrete points) well at low conversions (<20%). At higher conversions, however, the model tended to under predict methylpyrene reactivity slightly. This lack of quantitative agreement at higher conversions could be due to the model's omission of one or more secondary reactions. For example, pyrene could serve as a hydrogen acceptor in MD reactions, but the model does not include such steps. If it had, the predicted reactivity of methylpyrene would have increased. Of course, this lack of quantitative agreement between the model and the experiments may simply be due to the uncertainties in both the Arrhenius parameter estimates and the experimental data. Overall, however, the agreement between the calculated molar yields and the experimental molar yields is good at all three pyrolysis temperatures.

This favorable comparison allows us to use the model to explore the mechanistic scenarios and obtain information about the products for which we were not able to analyze (e.g., gases and high molecular weight compounds). For example, at 450°C the model predicts that the highest molar yield of 1,2-dipyrenylethane, the prominent termination product, should be 0.06 at 300 minutes. We did not observe this product, possibly because it did not elute from the GC column. The major gases predicted by the model were methane and hydrogen and the ratio of hydrogen to methane ranged from 0.14 to 0.25 over the temperature range studied (i.e., 400–450°C). This quantity is markedly lower than the value obtained experimentally from toluene pyrolysis (e.g., 1.5–2.3) over a temperature range of 640–695°C (Poutsma, 1990). The presence and relative amounts of these gases yields mechanistic insight into the pyrolysis of methylpyrene because H_2 is formed only through hydrogen abstraction by H-atoms, and methane is formed only after the hydrogenolytic displacement of the methyl group in methylpyrene. Because the rates of hydrogen abstraction by and addition of H-atoms are comparable in methylpyrene pyrolysis this low production of H_2 via abstraction suggests that H-atom ipso substitution plays a minor role in the hydrogenolysis of the methyl group. Furthermore, the low production of H_2 coupled with the high extent of hydrogenolysis suggests that the hydrogenolysis mechanism responsible for aryl-alkyl C–C bond cleavage in methylpyrene is highly efficient in its use of hydrogen.

Figure 11 displays the models' prediction of the rates of MD (Reaction 2), RHT (Reactions 4 and 5) and H-atom substitution (Reaction 8) which add hydrogen to the ipso position in methylpyrene as a function of batch holding time for the pyrolysis of methylpyrene at 425°C. The results at 400 and 450°C were similar to those presented in Figure 11, and the model predicts that the relative importance of these steps is a function of temperature and methylpyrene conversion. It is apparent from Figure 6 that the rate of H-atom ipso substitution is small in comparison with the rates of RHT and MD. The rate of RHT by hydropyrenyl radicals is initially faster than the rate of MD but at a conversion of about 0.1 the rate of MD becomes more significant. The model

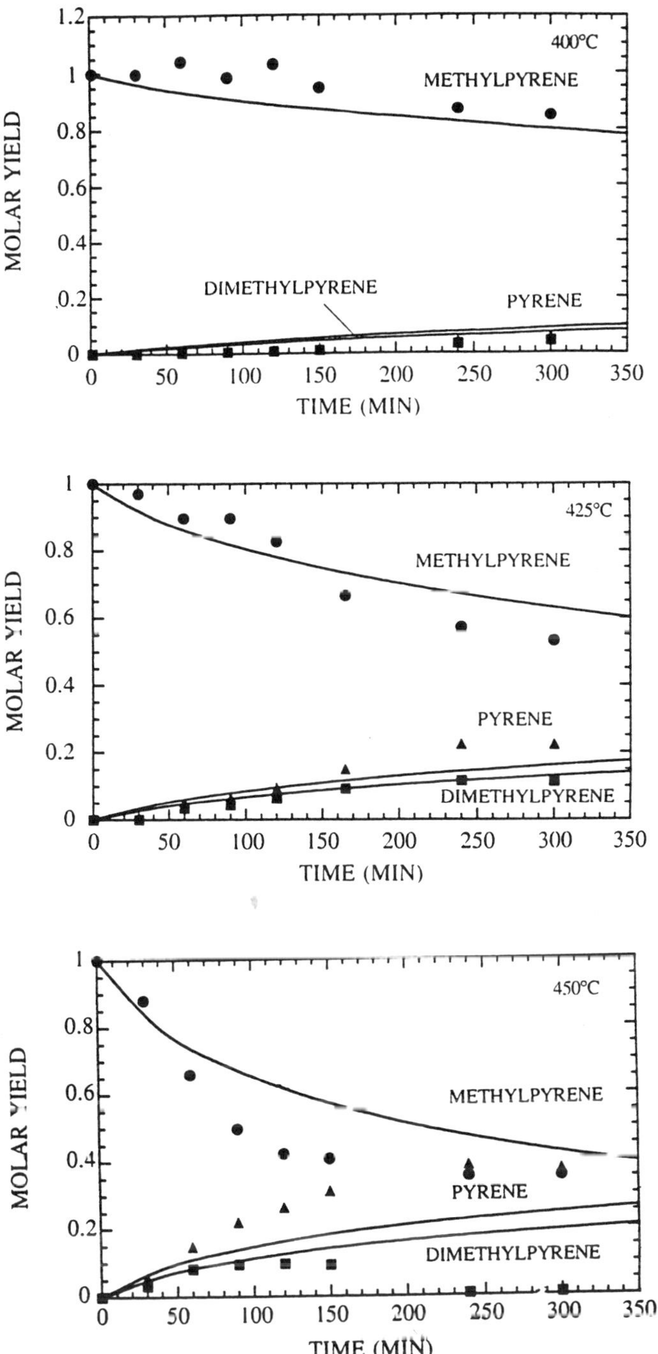

Figure 10. Comparison of Model and Experimental Results for Methylpyrene Pyrolysis (a) 400°C (b) 425°C (c) 450°C.

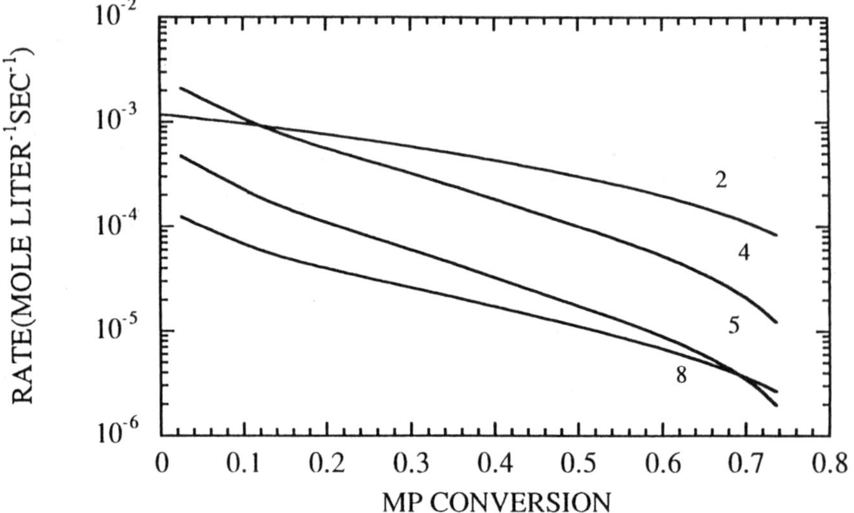

Figure 11. Comparison of Calculated Hydrogenolysis Rates for Methylpyrene Pyrolysis at 425°C.

predicted that this transition point occurred at methylpyrene conversion of about 0.2 and less than 0.1 for pyrolysis conditions of 400°C and 450°C respectively. Thus, the importance of RHT by hydropyrenyl radicals decreases as temperature increases. We expect that this behavior between these two step occurs because initially there is a very high concentration of hydropyrenyl radicals due to MD to non-ipso positions in methylpyrene. For example, disproportionation reactions decrease in the pool of hydropyrenyl radicals while regenerating methylpyrene molecules. Hence, the rate by which MD declines is slower than that of RHT.

The rates of RHT by methylhydropyrenyl radicals (Step 5) and H-atom substitution (Step 8) are lower than those of steps 2 and 4. At 425°C, the rate of RHT by methylcyclohexadienyl radicals is greater than the rate of H-atom substitution for methylpyrene conversions less than about 0.68. At conversions greater than 0.68, there is a shift in the importance of these two steps as the rate of Step 8 becomes greater than the rate of step 5. This behavior was also observed at the other temperatures studied. For example, at 400 and 450°C, this transition is predicted to occur at a methylpyrene conversion of 0.85 and 0.48 respectively. This trend suggests that the rate of Step 8 becomes relatively more important than the rate of Step 5 at higher temperatures.

The results of the pyrolysis simulation have revealed that the importance of step 4 decreases with increases in temperature and the role of H-atoms increases with increases with temperature. This behavior is due to the lower relative concentration of hydropyrenyl radicals at higher temperature. At higher temperatures, hydropyrenyl radicals are more likely to undergo β-scission to yield a hydrogen atom because of the high activation energy for this step. At higher temperatures there is a greater concentration of hydrogen atoms and thus a higher rate of H-atom ipso substitution.

The results of this mechanistic modelling also suggest that selective mechanisms (e.g., MD and RHT) play an important role in the cleavage of aliphatic substituents from aromatic rings, which is consistent with our experimental results. MD provides the means to either add hydrogen to the ipso-position or to generate hydrogen through the additition to non-ipso positions on the aromatic nucleus. In this latter mode, the aromatic nucleus accepts hydrogen and then donates hydrogen through RHT and regenerates the aromatic position which enables it to accept hydrogen again. Thus, the positions upon an aromatic nucleus are similar to the active sites upon a catalyst since they act as a reaction intermediate which intervenes between the reactants and products.

CONCLUSIONS

1. The pyrolysis of alkyl-substituted polycyclic aromatic compounds proceeds through three parallel pathways. The first of the two major pathways leads to three product lumps that are analogous to those observed for alkylbenzene pyrolysis. The second major pathway leads to products through the cleav-

age of the strong aryl-alkyl C–C bond. The third pathway, which is a minor one, leads to hydrogen deficient products.

2. The importance of aryl-alkyl C–C bond cleavage during the pyrolysis of polycyclic alkylaromatics depends primarily upon the localization energy at the specific point of substitution. The occurrence of aryl-alkyl C–C bond cleavage was independent of the number of rings, the length of the alkyl chain, and the type of condensation of the aromatic nucleus.

3. A quantitative correlation exists between the Dewar reactivity number, which can be easily calculated from perturbation molecular orbital theory, and the rate of aryl-alkyl C–C bond cleavage during the pyrolysis of n-alkylarenes. This correlation provides a link between the structure and the reactivity of polycyclic n-alkylaromatics and suggests that the mechanism responsible for the aryl-alkyl C–C bond cleavage is moderately selective. Further experimental evidence using 1,6 dimethylnaphthalene, a molecular probe, suggests that a selective hydrogenolysis mechanism is at work.

4. Mechanistic modeling of methylpyrene revealed that MD played an important role in generating hydropyrenyl radicals which then participated in radical hydrogen transfer. In the pyrolysis of methylpyrene, hydropyrenyl radicals generated from methyl radical addition were the key hydrogenolysis agents.

LITERATURE CITED

Ali, L. H.; Al-Ghannam, K. A. and Al-Rawi, J. M. "Chemical Structure of Asphaltenes in Heavy Crudes Investigated by n.m.r." *Fuel*, **1990**, (69), 519–521.

Altschuler, L.; Berliner, E. "Rates of Bromination of Polynuclear Aromatic Hydrocarbons" *J. Am. Chem. Soc.* **1966**, (88), 5837–5845.

Bhore, N. A.; Klein, M. T.; Bischoff, K. B. "The Delplot Technique: A New Method for Reaction Pathway Analysis" *Ind. Eng. Chem. Res.* **1990**, 29, 313–316.

Billmers, R.; Brown, R. L.; Stein, S. E. "Hydrogen Transfer Between 9,10 Dihydrophenathrene and Anthracene" *Int. J. Chem. Kinet.*. **1989**, 21, 375–389.

Billmers, R.; Griffith, L. L.; Stein, S. E. "Hydrogen Transfer Between Anthracene Structures" *J. Phys. Chem.* **1986**, 90, 517–523.

Bixel, J. C.; Lavaun, M. C.; Allered, V. D.; Benham, A. L. "Kinetics of Thermal Dealkylation of Alkylnaphthalenes". *Ind. Eng. Chem. Proc. Des. Dev.* **1964**, 3, 78–84.

Blouri, B.; Hamdan, F.; Herault, D. "Mild Cracking of High-Molecular-Weight Hydrocarbons". *Ind. Eng. Chem. Proc. Des. Dev.* **1985**, 24, 30–37.

Bockrath, B.C.; Schroeder, K.T.; Smith, M.R. "Investigation of Liquefaction Mechanisms with Molecular Probes". *Energy and Fuels* **1989**, 3, 268–272.

Boudart, M. *"Kinetics of Chemical Processes"* Prentice-Hall, Englewood, Cliffs, New Jersey, **1968**

Braun, W.; Herron, J. T.; Kahanen, D. K. "A Computer Program for Modelling Complex Chemical Reaction Systems" *Int. J. Chem. Kinet.* **1988**, *20*, 51–62.

Church, D. F.; Gleicher, G. J. "Addition of Vinylphenol to Polycyclic Vinylarenes" *J. Org. Chem.*,**1976**, *41*, 2327–2331.

Dewar, M. J. S.; Dougherty, R. C. *"The PMO Theory of Organic Chemistry"* Plenum Press, New York, **1975**.

Dewar, M. J. S. *"The Molecular Orbital Theory of Organic Chemistry"* McGraw Hill, New York, **1969**.

Dewar, M. J. S. "Molecular Orbital Theory of Organic Chemistry VI. Aromatic Substitution and Addition". *J. Am. Chem. Soc.* **1952**, *(74)*, 3357–3363.

Dewar, M. J. S.; Thompson, C. C. "Ground States of Conjugated Molecules IV. Estimation of Chemical Reactivity". *J. Am. Chem. Soc.* **1965**, *(87)*, 4414–4423.

Dickerman, S. C.; Feigenbaum, W. M.; Fryd, M.; Milstein, N.; Vermont, G. B.; Zimmerman, I.; McOmie, J.F.W. "Homolytic Aromatic Substitution. VIII Phenylation of Polycyclic Aromatic Hydrocarbons" *J. Am. Chem. Soc.* **1973**, *(95)*,4624–4631.

El-Mohamed, S.; Achard, M; Hardouin, F; Gasparoux, H. "Correlation between Diamagnetic Properties and Structural Characters of Asphaltenes and Other Heavy Petroleum Products". *Fuel* **1986**, *65*, 1501–1504.

Freund, H.; Matturro, M. G.; Olmsted, W. N.; Reynolds, R. P.; Upton, T. H. "Anomalous Side Chain Cleavage in Alkylaromatic Pyrolysis" *Prepr. Pap.-Am. Chem. Soc., Div. Fuel Chem.* **1990**, *35*, 496–504.

Freund, H.; Olmsted, W. N. " Detailed Kinetic Modeling of Butylbenzene Pyrolysis" *J. Int. Chem. Kinet.* **1989**, *21*, 561–574.

Gore, P. H.. "A Prediction of the Reactivity of Some Pent- and Hexa-cyclic Hydrocarbons" *J.Chem. Soc..* **1954**, 3166–3168.

Klein, M. T.; Virk, P. S. "Model Pathways in Lignin Thermolysis 1. Phenethyl Phenyl Ether" *Ind. Eng. Chem. Fundam.* **1983**, *22*, 35–45.

Malhotra, R.; McMillen, D. F, "A Mechanistic Numerical Model for Coal Liquefaction involving Hydrogenolysis of Strong Bonds. Rationalization of Interactive Effects of Solvent Aromaticity and Hydrogen Pressure". *Energy and Fuels* **1990**, *4*, 184–193.

McMillen, D. F, Golden, D. M. "Hydrocarbon Bond Dissociation Energies". *Ann. Rev. Phys. Chem.*, **1982**, *33*, 493–532.

McMillen, D. F.; Malhorta, R.; Chang, S.J.; Olgier, W.C.; Nigenda, E; Fleming, R. H. "Mechanisms of Hydrogen Transfer and Bond Scission of Strongly Bonded Coal Structures in Donor-Solvent Systems". *Fuel* **1987**, *66*, 1611–1620.

Mieville, R. L.; Trauth, D. M.; Robinson, K. K. "Asphaltene Characterization and Diffusion Measurements". *Prepr.-Am. Chem. Soc., Div. Pet. Chem.* **1989**, *34*, 635–643.

Mushrush, G. W.; Hazlett, R.N. "Pyrolysis of Organic Compounds Containing Long Unbranched Alkyl Groups". *Ind. Eng. Chem. Fundam.* **1984**, *23*, 288–294.

Poutsma, M. L. "Free-Radical Thermolysis and Hydrogenolysis of Model Hydrocarbons Relevant to Processing of Coal". *Energy and Fuels* **1990**, *4*, 113–131.

Savage, P.E.; Klein, M.T. "Kinetics of Coupled Reactions: Lumping Pentadecylbenzene Pyrolysis into Three Parallel Chains". *Chem. Eng. Sci.* **1989b**, *44*, 985–991.

Savage, P. E.; Klein, M.T. "Asphaltene Reaction Pathways. 2. Pyrolysis of *n*-Pentadecylbenzene". *Ind. Eng. Chem. Res.* **1987a**, *26*, 488–494.

Savage, P. E.; Klein, M.T. "Discrimination Between Molecular and Free-Radical Models of 1-Phenyldodecane Pyrolysis". *Ind. Eng. Chem. Res.* **1987b**, *26*, 374–376.

Savage, P.E.; Jacobs, G.E.; Javanmardian, M. "Autocatalysis and Aryl-Alkyl Bond Cleavage in 1-Dodecylpyrene Pyrolysis". *Ind. Eng. Chem. Res.* **1989**, *28*, 645–652.

Smith, C. M.; Savage, P. E. "Reactions of Polycyclic Alkylaromatics 1. Pathways, Kinetics, and Mechanisms for 1-Dodecylpyrene Pyrolysis". *Ind. Eng. Chem. Res.* **1991a,** *30*, 331–339.

Smith, C. M.; Savage, P. E. "Reactions of Polycyclic Alkylaromatics 2. Pyrolysis of 1,3 Diarylpropanes". *Energy and Fuels,.* **1991b**, *30*, 146–155.

Smith, C. M.; Savage, P. E. "Reactions of Polycyclic Alkylaromatics 3. Structure and Reactivity". *Submitted to AICHE Journal* **1991c**.

Smith, C. M.; Savage, P. E. "Reactions of Polycyclic Alkylaromatics 4. Hydrogenolysis Mechanisms in Alkylpyrene Pyrolysis". *Submitted to Energy and Fuels*.

Speight, J. G. In *Polynuclear Aromatic Compounds*; Advances in Chemistry Series No. 217; American Chemical Society: Washington, D.C. **1988**; 201–215.

Speight, J. G. "Latest Thoughts on the Molecular Nature of Petroleum Asphaltenes". *Prepr.-Am. Chem. Soc., Div. Pet. Chem.* **1989**, *34*, 321–328.

Streitwieser, A. " *Molecular Orbital Theory for Organic Chemists*" John Wiley and Sons, New York, **1961**.

Train, P. M.; Klein, M. T. In *Pyrolysis Oils from Biomass Producing, Analyzing, and Upgrading*; Advances in Chemistry Series No. 376; American Chemical Society: Washington, D.C. **1987**; pp.241–263.

Vernon, L. W. "Free Radical Chemistry of Coal Liquefaction: Role of Molecular Hydrogen". *Fuel* **1980**, *59*, 102–107.

Waller, P. R.; Williams, A.; Bartle, K.D. "The Structural Nature and Solubility of Residual Fuel Oil Fractions". *Fuel* **1989**, *68*, 520–526.